Advances in Experimental Medicine and Biology

Volume 1218

Advances in Experimental Medicine and Biology provides a platform for scientific contributions in the main disciplines of the biomedicine and the life sciences. This book series publishes thematic volumes on contemporary research in the areas of microbiology, immunology, neurosciences, biochemistry, biomedical engineering, genetics, physiology, and cancer research. Covering emerging topics and techniques in basic and clinical science, it brings together clinicians and researchers from various fields

Advances in Experimental Medicine and Biology has been publishing exceptional works in the field for over 40 years, and is indexed in SCOPUS, Medline (PubMed), Journal Citation Reports/Science Edition, Science Citation Index Expanded (SciSearch, Web of Science), EMBASE, BIOSIS, Reaxys, EMBiology, the Chemical Abstracts Service (CAS), and Pathway Studio.

2018 Impact Factor: 2.126.

More information about this series at http://www.springer.com/series/5584

Jörg Reichrath • Sandra Reichrath
Editors

Notch Signaling in Embryology and Cancer

Notch Signaling in Embryology

Editors
Jörg Reichrath
Department of Dermatology
The Saarland University Hospital
Homburg, Saarland, Germany

Sandra Reichrath
Department of Dermatology
The Saarland University Hospital
Homburg, Saarland, Germany

ISSN 0065-2598 ISSN 2214-8019 (electronic)
Advances in Experimental Medicine and Biology
ISBN 978-3-030-34438-2 ISBN 978-3-030-34436-8 (eBook)
https://doi.org/10.1007/978-3-030-34436-8

This Springer imprint is published by the registered company Springer Nature Switzerland AG
The registered company address is: Gewerbestrasse 11, 6330 Cham, Switzerland

We dedicate this book to our sons, Niklas and Benjamin, the absolute joy of our lives.

Preface

When the American scientist John S. Dexter discovered mutant fruit flies (*Drosophila melanogaster*) in his research laboratory at Olivet College (Olivet, Michigan) more than a century ago, he could not have expected the tremendous impact that the characteristic notch-wing phenotype (a nick or notch in the wingtip that gave the responsible gene the name Notch) would later have for many fields of biology and medicine, including embryology, genetics, and cancer. During the last decades, a huge mountain of impressive scientific process has convincingly demonstrated that Notch signaling represents one of the most fascinating pathways that govern cellular core processes including cell fate decisions, embryogenesis, and adult tissue homeostasis. Therefore, it is no surprise that the first edition of the book *Notch Signaling in Embryology and Cancer*, published by Landes and Springer in 2012 in the prestigious series Advances in Experimental Medicine and Biology, was very successful, for it fulfilled the need to provide a broad audience (ranging from medical students to basic scientists, physicians, and all other healthcare professionals) with up-to-date information in a comprehensive, highly readable format. At this time, it was the benchmark on this topic, with individual chapters being written by highly respected experts in the field. Because of the enormous progress that has been made on this topic in recent years, we have decided that it is now the right time to publish an updated and extended version. The second edition of the abovementioned book has been expanded substantially and consequently has been divided in three separate volumes to include many new chapters. In its different volumes, leading experts in the field present a comprehensive, highly readable overview on selected aspects of three important topics related to Notch signaling, namely, the underlying molecular mechanisms that mediate its biological effects (Volume I), its role in embryogenesis (Volume II), and last but not least its relevance for pathogenesis, progression, prevention, and therapy of cancer (Volume III). This second volume summarizes the fascinating role of this pathway, which was first developed during the evolution in metazoans and was first discovered in the fruit fly *Drosophila melanogaster*, for embryology. We are convinced that it will be as successful as the previous edition and are very grateful for the willingness of all authors to contribute to this book. We would also like to express our thanks to Murugesan Tamilselvan, Anthony Dunlap, Larissa Albright, and all other members of the Springer staff for their expertise, diligence, and patience in helping us complete this book.

Enjoy the reading!

Homburg, Saarland, Germany

Jörg Reichrath
Sandra Reichrath

Contents

Contributors

Shahrzad Bahrampour The Hospital for Sick Children, Peter Gilgan Center for Research and Learning, Toronto, ON, Canada

Karolinska Institutet, Department of Cell and Molecular Biology (CMB), Stockholm, Sweden

Sergio Córdoba Centro de Biología Molecular "Severo Ochoa", CSIC-UAM, Madrid, Spain

Department of Biology, New York University, New York, NY, USA

Nicolas Daudet University College London, The Ear Institute, London, UK

Carlos Estella Centro de Biología Molecular "Severo Ochoa", CSIC-UAM, Madrid, Spain

Renjie Jiao Sino-French Hoffmann Institute, School of Basic Medical Sciences, Guangzhou Medical University, Guangzhou, China

Chung-Weng Phang Sino-French Hoffmann Institute, School of Basic Medical Sciences, Guangzhou Medical University, Guangzhou, China

Jörg Reichrath Department of Dermatology, The Saarland University Hospital, Homburg, Germany

Sandra Reichrath Department of Dermatology, The Saarland University Hospital, Homburg, Germany

Tetsuichiro Saito Department of Developmental Biology, Graduate School of Medicine, Chiba University, Chiba, Japan

Makoto Sato Mathematical Neuroscience Unit, Institute for Frontier Science Initiative, Kanazawa University, Kanazawa-shi, Ishikawa, Japan

Laboratory of Developmental Neurobiology, Graduate School of Medical Sciences, Kanazawa University, Kanazawa-shi, Ishikawa, Japan

Stefan Thor School of Biomedical Sciences, University of Queensland, St Lucia, QLD, Australia

Chuanxian Wei Sino-French Hoffmann Institute, School of Basic Medical Sciences, Guangzhou Medical University, Guangzhou, China

Tetsuo Yasugi Mathematical Neuroscience Unit, Institute for Frontier Science Initiative, Kanazawa University, Kanazawa-shi, Ishikawa, Japan

Magdalena Żak University College London, The Ear Institute, London, UK

About the Editors

Jörg Reichrath and **Sandra Reichrath** are working at the Clinic for Dermatology, Allergology, and Venerology of the Saarland University Hospital in Homburg/Saar, Germany. Their main research interests include photobiology, carcinogenesis of skin cancer (with a special interest for the role of ultraviolet radiation, Notch, and p53), and the relevance of the vitamin D endocrine system for health and disease. Sandra received her academic degrees (Dipl. Biol. and Dr. Rer. Nat.) from the Albert-Ludwigs-Universität (Freiburg, Germany) and from Saarland University (Saarbrücken/Homburg, Germany), respectively, while Jörg received his academic degrees (Dr. Med., Venia Legendi) from Saarland University (Saarbrücken/Homburg, Germany). He has been awarded numerous prizes including the Arnold Rikli Prize in 2006.

Chapter 1
Notch Signaling and Tissue Patterning in Embryology: An Introduction

Jörg Reichrath and Sandra Reichrath

Abstract The attention of science first turned to the gene that later earned the name *Notch* over a century ago, when the American scientist John S. Dexter discovered in his laboratory at Olivet College the characteristic notched-wing phenotype (a nick or notch in the wingtip) in mutant fruit flies *Drosophila melanogaster*. At present, it is generally accepted that the Notch pathway governs tissue patterning and many key cell fate decisions and other core processes during embryonic development and in adult tissues. Not surprisingly, a broad variety of independent inherited diseases (including CADASIL, Alagille, Adams-Oliver, and Hajdu-Cheney syndromes) have now convincingly been linked to defective Notch signaling. In the second edition of the book entitled *Notch Signaling in Embryology and Cancer*, leading researchers provide a comprehensive, highly readable overview on molecular mechanisms of Notch signaling (Volume I), and notch's roles in embryology (Vol. II) and cancer (Vol. III). In these introductory pages of Vol. II, we give a short overview on its individual chapters, which are intended to provide both basic scientists and clinicians who seek today's clearest understanding of the broad role of Notch signaling in embryology with an authoritative day-to-day source.

Keywords Notch · Notch signaling · Notch pathway · Embryonic development · Jagged · Delta-like ligand · Cell fate decisions · Tissue patterning

It is now generally accepted that, from sponges, roundworms, *Drosophila melanogaster*, and mice to humans, the Notch pathway governs tissue patterning and many key cell fate decisions and other core processes during embryonic development and in adult tissues (Andersson et al. 2011). When the first edition of *Notch Signaling in Embryology and Cancer* was published by Landes and Springer in 2012 in the

---

J. Reichrath (✉) · S. Reichrath
Department of Dermatology, The Saarland University Hospital, Homburg, Germany
e-mail: joerg.reichrath@uks.eu

J. Reichrath, S. Reichrath (eds.), *Notch Signaling in Embryology and Cancer*,
Advances in Experimental Medicine and Biology 1218,
https://doi.org/10.1007/978-3-030-34436-8_1

prestigious series *Advances in Experimental Medicine and Biology*, it was the benchmark on this topic, providing a broad audience (ranging from medical students to basic scientists, physicians, and all other health-care professionals) with up-to-date information in a comprehensive, highly readable format. As the result of the huge mountain of new scientific findings that has been build up in the meantime, which underlines the high biological/clinical relevance of Notch signaling and further unravels their underlying molecular mechanisms, we have decided that it is now the right time to publish an updated and extended version. The second edition of this book has been expanded substantially and has been divided in three separate volumes to include many new chapters. In the different volumes of this book, leading researchers provide a comprehensive, highly readable overview on three important topics related to Notch signaling, namely, the underlying molecular mechanisms that mediate its biological effects (volume I), its role in embryonic development (volume II), and finally its relevance for pathogenesis, progression, prevention, and therapy of cancer (volume III). This second volume summarizes the role of the Notch pathway, which first developed during evolution in metazoans (Gazave et al. 2009; Richards and Degnan 2009) and that was first discovered in a fruit fly (*Drosophila melanogaster*), for tissue patterning and embryonic development. As outlined elsewhere in this book (Reichrath and Reichrath 2020a), the tale that created the name *Notch* began over a century ago, when the American scientist John S. Dexter discovered in his laboratory at Olivet College (Olivet, Michigan, USA) the characteristic notched-wing phenotype (a nick or notch in the wingtip) in mutant fruit flies *Drosophila melanogaster* (Dexter 1914). The alleles causing this phenotype were identified 3 years later at Columbia University (New York City, New York, USA) by another American scientist, Thomas Hunt Morgan (1866–1945) (Morgan 1917), who discovered various mutant loci in the chromosomes of these fruit flies that were associated with several distinct notched-wing phenotypes. Although the majority of them were lethal, these alleles were associated with the characteristic phenotype (a nick in the wingtip and bristle phenotype specifically in female fruit flies), suggesting an association of these alleles with the X chromosome (Morgan 1928). Notably, this discovery and similar investigations that supported the chromosomal theory of inheritance earned Thomas Hunt Morgan in 1933 the Nobel Prize in physiology/medicine. In subsequent decades, despite the extensive research on the *Notch* locus, researchers struggled to identify the function for the *Notch* gene due to the lethality early in embryogenesis and the broad variety of phenotypic consequences of Notch mutants. In the following years, many additional alleles were identified, which were associated with the Notch phenotype. These observations were finally confirmed by cloning and sequencing of the mutant *Notch* locus in the laboratories of Spyros Artavanis-Tsakonas and Michael W. Young, more than half a century later (Wharton et al. 1985; Kidd et al. 1986).

Moreover, a broad variety of independent inherited diseases linked to defective Notch signaling has now been identified, highlighting its clinical relevance. The discovery of these congenital diseases started in 1996 in patients diagnosed with CADASIL (cerebral autosomal dominant arteriopathy with subcortical infarcts and leukoencephalopathy, an autosomal dominant hereditary stroke disorder resulting in vascular dementia) (Joutel et al. 1996), with the linkage analysis-based discovery of heterozygous *NOTCH3* mutations on chromosome 19. In the next year, two laboratories published independently the identification of *JAG1* as the gene within chromosome 20p12 that causes Alagille syndrome (Li et al. 1997; Oda et al. 1997). Since these pioneer investigations, several additional inherited disorders, including Adams-Oliver and Hajdu-Cheney syndromes, have now convincingly been linked to defective Notch signaling. Many of these congenital diseases are rare (prevalences of just a few cases per 100,000), presenting on the one hand severe hurdles to investigating the impact of these genes in humans but demonstrating on the other hand how important Notch pathway components are for human survival. Fortunately, the generation and investigation of knockout mice and other animal models have in recent years resulted in a huge mountain of new information concerning Notch gene function, allowing to separate the role of specific Notch components in human development and disease.

This volume is intended to provide both basic scientists and clinicians who seek today's clearest understanding of the broad role of Notch signaling in embryology with an authoritative day-to-day source. In the first chapter following this introduction, Reichrath and Reichrath give a short overview on the role of Notch signaling for the embryonic development of several selected tissues, namely, the brain, skin, kidneys, liver, pancreas, sensory organs, skeleton, heart, and vascular system (Reichrath and Reichrath 2020a).

In the following chapter, Shahrzad Bahrampour and Stefan Thor discuss the impact of Notch signaling for brain development in detail (Bahrampour and Thor 2020). They point out that, during central nervous system (CNS) development, a complex series of events play out, starting with the establishment of neural progenitor cells, followed by their asymmetric division and formation of lineages and the differentiation of neurons and glia. Studies in the *Drosophila melanogaster* embryonic CNS have revealed that the Notch signal transduction pathway plays at least five different and distinct roles during these events. In their chapter, Bahrampour and Thor review these many faces of Notch signaling and discuss the mechanisms that ensure context-dependent and compartment-dependent signaling. The authors conclude by discussing some outstanding issues regarding Notch signaling in this system, which likely have bearing on Notch signaling in many species.

In the next chapter Wei, Phang, and Jiao underline that the simplicity of the Notch pathway in *Drosophila melanogaster*, in combination with the availability of powerful genetics, makes it an attractive model for studying the fundamental mechanisms of how Notch signaling is regulated and how it functions in various cellular conditions during embryonic development (Wei et al. 2020). In this context, the authors summarize the research advances in *Drosophila* development on the epigenetic mechanisms by which the chromatin assembly factor-1 (CAF-1) regulates Notch signaling activity, which enables Notch to orchestrate different biological inputs and outputs in specific cellular contexts. They convincingly demonstrate that epigenetic regulation of Notch signaling by CAF-1 and other epigenetic regulators plays essential roles in fine-tuning the transcriptional output of Notch signaling to coordinate multicellular organism development. The authors conclude that it remains an open question as to why and how different epigenetic regulators are involved in mediating different histone modifications status, leading to different transcriptional outputs of either gene repression or gene activation in one specific signal transduction pathway.

Underlining the many facettes of Notch signaling for embryonic development, Makoto Sato and Tetsuo Yasugi discuss in the following chapter the relevance of a combination of Notch-mediated lateral inhibition and epidermal growth factor (EGF)-mediated reaction diffusion for the regulation of proneural wave propagation (Sato and Yasugi 2020). They report that during various biological processes, Notch has to act together with other signaling systems to regulate binary cell fate choice via lateral inhibition resulting in salt-and-pepper pattern formation. However, they emphasize that it is in many cases not clear what happens when Notch is combined with other signaling systems and that mathematical modelling and the use of a simple biological model system will be essential to address this uncertainty. They explain that a wave of differentiation in the *Drosophila* visual center, the "proneural wave," accompanies the activity of the Notch and EGF signaling pathways and that, although all of the Notch signaling components required for lateral inhibition are involved in the proneural wave, no salt-and-pepper pattern is found during the progression of the proneural wave. Instead, Notch is activated along the wave front and regulates proneural wave progression. Makoto Sato and Tetsuo Yasugi ask the question how does Notch signaling control wave propagation without forming a salt-and-pepper pattern? As they point out, a mathematical model of the proneural wave based on biological evidence convincingly demonstrated that Notch-mediated lateral inhibition is implemented within the proneural wave and that the diffusible action of EGF cancels salt-and-pepper pattern formation. They discuss that the results from numerical simulation have been confirmed by genetic experiments in vivo and suggest that the combination of Notch-mediated lateral inhibition and EGF-mediated reaction diffusion enables a novel function of Notch signaling that regulates propagation of the proneural wave. Makoto Sato and Tetsuo Yasugi conclude that similar mechanisms may play important roles in diverse biological processes found in embryonic development and cancer pathogenesis.

In the following chapter, Tetsuichiro Saito convincingly demonstrates that a nucleolar protein, Nepro, is essential for the maintenance of early neural stem cells and preimplantation embryos (Saito 2020). He points out that Notch signaling is required for maintaining neural stem cells (NSCs) in the developing brain and that NSCs have potential to give rise to many neuronal types in the early telencephalon, and the potential decreases as embryonic development proceeds. Tetsuichiro Saito explains that *Nepro*, which encodes a unique nucleolar protein and is activated downstream of Notch, is essential for maintaining NSCs in the early telencephalon. *Nepro* is also expressed at basal levels and required for maintaining the preimplantation embryo, by repressing mitochondria-associated p53 apoptotic signaling. Tetsuichiro Saito points out that Notch signaling also controls dendritic complexity in mitral cells, major projection neurons in the olfactory bulb, and concludes that many steps of neural development involve Notch signaling.

In the following chapter, Sergio Córdoba and Carlos Estella (2020) summarize our present understanding of the role of Notch signaling for leg development in *Drosophila melanogaster*. They explain that the Notch pathway plays diverse and fundamental roles during animal development. One of the most relevant, which arises directly from its unique mode of activation, is the specification of cell fates and tissue boundaries. The development of the leg of *Drosophila melanogaster* is a fine example of this Notch function, as it is required to specify the fate of the cells that will eventually form the leg joints, the flexible structures that separate the different segments of the adult leg. Notch activity is accurately activated and maintained at the distal end of each segment in response to the proximo-distal patterning gene network of the developing leg. Region-specific downstream targets of Notch in turn regulate the formation of the different types of joints. The authors discuss recent findings that shed light on the molecular and cellular mechanisms that are ultimately governed by Notch to achieve epithelial fold and joint morphogenesis. Finally, they briefly summarize the role that Notch plays in inducing the nonautonomous growth of the leg. Overall, this book chapter aims to highlight leg development as a useful model to study how patterning information is translated into specific cell behaviors that shape the final form of an adult organ.

In the next chapter, Nicolas Daudet and Magdalena Żak explain the role of Notch signalling as the multitask manager of inner ear development and regeneration (Daudet and Żak 2020). They point out that Notch signalling is a major regulator of tissue patterning in metazoans, exerting its effects both by lateral inhibition (whereby Notch mediates competitive interactions between cells to limit adoption of a given developmental fate) and by lateral induction (a cooperative mode of action that was originally described during the patterning of the Drosophila wing disc and creates boundaries or domains of cells of the same character). In their chapter, Nicolas Daudet and Magdalena Żak introduce these two signalling modes and explain how they contribute to distinct aspects of the development and regeneration of the vertebrate inner ear, the organ responsible for the perception of sound and head movements. Moreover, Nicolas Daudet and Magdalena Żak discuss in this chapter some of the factors that influence the context-specific outcomes of Notch signalling in the inner ear, and the ongoing efforts to target this pathway for the treatment of hearing loss and vestibular dysfunction.

In the last chapter, Reichrath and Reichrath give a short overview on inherited diseases related to defective Notch signaling, including CADASIL (cerebral autosomal dominant arteriopathy with subcortical infarcts and leukoencephalopathy) and Alagille, Adams-Oliver, Hajdu-Cheney, and lateral meningocele syndromes (Reichrath and Reichrath 2020b). They point out that the evolutionary highly conserved Notch pathway governs many cellular core processes including cell fate decisions, although it is characterized by a simple molecular design. Moreover, Notch signaling, which first developed in metazoans, represents one of the most important pathways that govern embryonic development. Consequently, a broad variety of independent inherited diseases linked to defective Notch signaling has now been identified, including Alagille, Adams-Oliver, and Hajdu-Cheney syndromes, CADASIL, early-onset arteriopathy with cavitating leukodystrophy, lateral meningocele syndrome (LMS), and infantile myofibromatosis. In their review, Reichrath and Reichrath give a brief overview on molecular pathology and clinical findings in congenital diseases linked to the Notch pathway (Reichrath and Reichrath 2020b). Moreover, they discuss the emerging role of Notch as a promising therapeutic target. In this context, it is of interest that in a mouse model of LMS (Notch3$^{tm1.1Ecan}$), cancellous bone osteopenia was no longer detected after intraperitoneal administration of antibodies directed against the negative regulatory region (NRR) of Notch3 (Yu et al. 2019, reviewed in Reichrath and Reichrath 2020b). In that study, anti-Notch3 NRR antibody suppressed expression of Hes1, Hey1, and Hey2 (Notch target genes) and decreased Tnfsf11 (receptor activator of NF kappa B ligand) messenger RNA in Notch3$^{tm1.1Ecan}$ osteoblast cultures (Yu et al. 2019, reviewed in Reichrath and Reichrath 2020b). This study indicates that cancellous bone osteopenia of Notch3$^{tm1.1Ecan}$ mutants can be reversed by anti-Notch3 NRR antibodies, thereby opening new avenues for treatment of bone osteopenia in LMS patients (Yu et al. 2019, reviewed in Reichrath and Reichrath 2020b).

We hope that this volume will provide both basic scientists and clinicians who seek today's clearest understanding of the broad and fascinating role of Notch signaling for the embryonic development with an authoritative day-to-day source.

References

Andersson ER, Sandberg R, Lendahl U (2011) Notch signaling: simplicity in design, versatility in function. Development 138:3593–3612. https://doi.org/10.1242/dev.063610

Bahrampour S, Thor S (2020) The five faces of Notch signalling during *Drosophila melanogaster* embryonic CNS development. Adv Exp Med Biol. 1218:39–58

Córdoba S, Estella C (2020) The role of Notch signaling in leg development in *Drosophila melanogaster*. Adv Exp Med Biol. 1218:103–128

Daudet N, Żak M (2020) Notch signalling: the multitask manager of inner ear development and regeneration. Adv Exp Med Biol. 1218:129–158

Dexter JS (1914) The analysis of a case of continuous variation in Drosophila by a study of its linkage relations. Am Nat 48:712–758. https://doi.org/10.1086/279446

Gazave E, Lapébie P, Richards GS, Brunet F, Ereskovsky AV, Degnan BM, Borchiellini C, Vervoort M, Renard E (2009) Origin and evolution of the Notch signalling pathway: an overview from eukaryotic genomes. BMC Evol Biol 9:249. https://doi.org/10.1186/1471-2148-9-249

Joutel A, Corpechot C, Ducros A, Vahedi K, Chabriat H, Mouton P, Alamowitch S, Domenga V, Cécillion M, Maréchal E et al (1996) Notch3 mutations in CADASIL, a hereditary adult-onset condition causing stroke and dementia. Nature 383:707–710. https://doi.org/10.1038/383707a0

Kidd S, Kelley MR, Young MW (1986) Sequence of the notch locus of Drosophila melanogaster: relationship of the encoded protein to mammalian clotting and growth factors. Mol Cell Biol 6(9):3094–3108

Li L, Krantz ID, Deng Y, Genin A, Banta AB, Collins CC, Qi M, Trask BJ, Kuo WL, Cochran J et al (1997) Alagille syndrome is caused by mutations in human Jagged1, which encodes a ligand for Notch1. Nat Genet 16:243–251. https://doi.org/10.1038/ng0797-243

Morgan TH (1917) The theory of the gene. Am Nat 19:309–310. https://doi.org/10.1086/279629

Morgan T (1928) The theory of the gene, revised edn. Yale University Press, New Haven, pp 77–81

Oda T, Elkahloun AG, Pike BL, Okajima K, Krantz ID, Genin A, Piccoli DA, Meltzer PS, Spinner NB, Collins FS et al (1997) Mutations in the human Jagged1 gene are responsible for Alagille syndrome. Nat Genet 16:235–242. https://doi.org/10.1038/ng0797-235

Reichrath J, Reichrath S (2020a) Notch signaling and embryonic development: an ancient friend, revisited. Adv Exp Med Biol. 1218:9–38

Reichrath J, Reichrath S (2020b) Notch pathway and inherited diseases: challenge and promise. Adv Exp Med Biol. 1218:159–188

Richards GS, Degnan BM (2009) The dawn of developmental signaling in the metazoa. Cold Spring Harb Symp Quant Biol 74:81–90. https://doi.org/10.1101/sqb.2009.74.028

Saito T (2020) A nucleolar protein, Nepro, is essential for the maintenance of early neural stem cells and preimplantation embryos. Adv Exp Med Biol 1218:93–102

Sato M, Yasugi T (2020) Regulation of proneural wave propagation through a combination of Notch-mediated lateral inhibition and EGF-mediated reaction diffusion. Adv Exp Med Biol 1218:77–92

Wei C, Phang C.-W, Jiao R (2020) Epigenetic regulation of Notch signaling during *Drosophila* development. Adv Exp Med Biol 1218:59–76

Wharton KA, Johansen KM, Xu T, Artavanis-Tsakonas S (1985) Nucleotide sequence from the neurogenic locus notch implies a gene product that shares homology with proteins containing EGF-like repeats. Cell 43(3 Pt 2):567–581

Yu J, Siebel CW, Schilling L, Canalis E (2019) An antibody to Notch3 reverses the skeletal phenotype of lateral meningocele syndrome in male mice. J Cell Physiol. https://doi.org/10.1002/jcp.28960. [Epub ahead of print] PubMed PMID: 31188489

Chapter 2
Notch Signaling and Embryonic Development: An Ancient Friend, Revisited

Jörg Reichrath and Sandra Reichrath

Abstract The evolutionary highly conserved Notch pathway, which first developed during evolution in metazoans and was first discovered in fruit flies (*Drosophila melanogaster*), governs many core processes including cell fate decisions during embryonic development. A huge mountain of scientific evidence convincingly demonstrates that Notch signaling represents one of the most important pathways that regulate embryogenesis from sponges, roundworms, *Drosophila melanogaster*, and mice to humans. In this review, we give a brief introduction on how Notch orchestrates the embryonic development of several selected tissues, summarizing some of the most relevant findings in the central nervous system, skin, kidneys, liver, pancreas, inner ear, eye, skeleton, heart, and vascular system.

Keywords Notch · Notch signaling · Notch pathway · Embryonic development · Jagged · Delta-like ligand

Abbreviations

BMP	Bone morphogenetic protein
cKO	Conditional knockout
CNS	Central nervous system
DG	Dentate gyrus
Dll	Delta-like
E	Embryonic day
FGF	Fibroblast growth factor
Hes	Hairy and enhancer of split
Hf	Hair follicle
IP	Intermediate progenitor

J. Reichrath (✉) · S. Reichrath
Department of Dermatology, The Saarland University Hospital, Homburg, Germany
e-mail: joerg.reichrath@uks.eu

J. Reichrath, S. Reichrath (eds.), *Notch Signaling in Embryology and Cancer*, Advances in Experimental Medicine and Biology 1218,
https://doi.org/10.1007/978-3-030-34436-8_2

IPC	Intermediate progenitor cell
KO	Knockout
LV	Lateral ventricle
LW	Lateral wall
MET	Mesenchymal-to-epithelial transition
NEPs	Neuroepithelial cells
NICD	Notch intracellular domain
NSCs	Neural stem cells
OB	Olfactory bulb
P	Postnatal day
RA	Retinoic acid
RGC	Radial glia cell
RMS	Rostral migratory stream
SGZ	Subgranular zone
Shh	Sonic hedgehog
SVZ	Subventricular zone
VZ	Ventricular zone
Wnt	Wingless

Introduction

There is now general consensus that Notch signaling, which is evolutionary highly conserved since it first developed in metazoans (Gazave et al. 2009; Richards and Degnan 2009) and was first discovered in the fruit fly *Drosophila melanogaster*, represents one of the most important cellular pathways that govern and orchestrate embryonic development. Notably, the Notch pathway, which is simple in design but has a striking versatility in function, regulates many key cell decisions and other core processes during embryogenesis from sponges, roundworms, *Drosophila melanogaster*, and mice to humans (Andersson et al. 2011). The tale that earned the gene the name *Notch* started over a century ago at Olivet College (Olivet, Michigan, USA), where at that time the American scientist John S. Dexter observed the characteristic notched-wing phenotype (a nick or notch in the wingtip) in his mutant fruit flies *Drosophila melanogaster* (Dexter 1914). The alleles that caused this phenotype were discovered 3 years later at Columbia University (New York City, New York, USA) by another American scientist, Thomas Hunt Morgan (September 25, 1866–December 4, 1945), who observed several mutant loci in the chromosomes of these fruit flies that were associated with several distinct notched-wing phenotypes (Morgan 1917). Although the majority of them were lethal, these alleles were associated with the characteristic phenotype with a nick in the wingtip and bristle phenotype specifically in female fruit flies, suggesting an association of these alleles with the X chromosome (Morgan 1928). The studies of Dexter (1914) and Morgan (1917, 1928) represent important steps in the transition of embryological research from traditional, largely descriptive morphology to an experimental embryology that sought physical and chemical explanations for organismal development. In the following years, many additional alleles were identified, which were associated with the Notch phenotype (Morgan 1928). Notably, these and related investigations that supported the chromosomal theory of inheritance earned Thomas Hunt Morgan in 1933 the Nobel Prize in Physiology/Medicine. Although the following decades were characterized by extensive research on the *Notch* locus, scientists had to overcome many difficulties to reach their goal to better understand the function of the *Notch* gene. These obstacles were due to the lethality early in embryogenesis and the broad phenotypic consequences of many Notch mutants. A milestone that finally confirmed the thoughtful observations that Thomas Hunt Morgan published in 1917 was the cloning and sequencing of the mutant *Notch* locus by the laboratories of Spyros Artavanis-Tsakonas and Michael W. Young more than half a century later (Wharton et al. 1985; Kidd et al. 1986). During the last decades, a huge mountain of scientific information has convincingly shown that *Notch* plays multiple roles both in embryogenesis and in adult tissue homeostasis. Although it is simple in design, the Notch pathway is extraordinarily versatile in function (Andersson et al. 2011). The flagship functions include many important cell fate decisions, such as keeping precursor and stem cells in a non-differentiated state, and the ability to activate and orchestrate cell proliferation. In general, these functions involve canonical, ligand-dependent Notch activation. However, ligand-independent Notch activation has also been described in several cellular contexts.

Until today, a huge mountain of studies – ranging from the elucidation of the Notch pathway (reviewed by Bray 2016; Kopan and Ilagan 2009) to the generation of knockouts in model organisms and the discovery of Notch genes mutated in humans (Gridley 2003) – has confirmed an essential role for Notch signaling in human development. Consequently, a variety of independent inherited diseases related to defective Notch signaling has been identified, underlining the relevance of Notch for embryogenesis. The discovery of these congenital diseases started in 1996, when a pioneer study reported the linkage analysis-based detection of heterozygous *NOTCH3* mutations on chromosome 19 in patients diagnosed with CADASIL (cerebral autosomal dominant arteriopathy with subcortical infarcts and leukoencephalopathy, an autosomal dominant hereditary stroke disorder resulting in vascular dementia) (Joutel et al. 1996). In the following year, two laboratories independently published that *JAG1* was the gene within chromosome 20p12 that was responsible for Alagille syndrome (Li et al. 1997; Oda et al. 1997). Since then, several other inherited disorders, involving pathological embryonic development of various tissues, including Adams-Oliver and Hajdu-Cheney syndromes, have now convincingly been linked to defective Notch signaling (reviewed in Reichrath and Reichrath 2019). In this review, we give a brief introduction on the role of Notch signaling for the embryonic development of several selected tissues, namely, the brain, skin, kidneys, liver, pancreas, sensory organs, skeleton, heart, and vascular system, and discuss the role of Notch as an emerging therapeutic target.

Challenge and Promise: The Impact of Notch Signaling for the Embryogenesis of the Central Nervous System (CNS)

Embryogenesis of the Central Nervous System (CNS): A Brief Introduction

The embryonic development of the brain and of other parts of the central nervous system (CNS) is tightly regulated in time and space by a complex network of many different signaling pathways, including the Notch pathway (reviewed in Engler et al. 2018). Notably, Notch signaling represents a key regulator of cell fate decisions, and maintenance mechanisms of neural stem cells (NSCs), which generate both in the developing embryonic and in the adult brain all neurons, are highly conserved during evolution and are present from *Drosophila melanogaster* to humans (Artavanis-Tsakonas et al. 1999; Kazanis et al. 2008; Fuentealba et al. 2015; reviewed in Engler et al. 2018). Notch signaling is linked to many different aspects of embryonic brain development, during which the generating of the neural plate, a process that is named neurulation, begins in mice around embryonic day 8 (E8) (reviewed in Engler et al. 2018). This tightly regulated process is induced via a well-balanced orchestration of stimulating factors and inhibiting signals acting on several anatomical structures, namely, the notochord, the dorsal ectoderm, and the Spemann organizer (Tam and Loebel 2007; reviewed in Engler et al. 2018). After the neural plate has been generated on the dorsal side of the embryo by neural ectodermal cells (radial progenitors spanning the thickness of the neural plate also named neuroepithelial cells (NEPs)), these cells begin to become regionalized, a process that has been linked to Notch signaling (reviewed in Engler et al. 2018). Notably, the NEPs at the lateral edges of the plate become the multipotent neural crest stem cells, the precursors of the peripheral nervous system, melanocytes, and, in some regions of the embryo, specific bones and muscles (Bhatt et al. 2013; Sauka-Spengler and Bronner 2010; reviewed in Engler et al. 2018). The neural plate then invaginates into the embryo at the midline and the lateral edges of the neural plate fold dorsally (reviewed in Engler et al. 2018). Thereafter, the two lateral edges of the neural plate meet and fuse at the dorsal midline, zippering closed in the anterior and posterior directions, a process that begins at the future hindbrain region (reviewed in Engler et al. 2018). The neural plate then generates the neural tube (reviewed in Engler et al. 2018). Notably, NEPs in the neuroepithelium of the neural tube represent the first stem cells of the central nervous system (reviewed in Engler et al. 2018).

It has convincingly been shown that NSCs are the origins of the neurogenic lineage (reviewed in Engler et al. 2018). In the developing brain, radial glia in the ventricular zone (VZ) are the stem cells (Greig et al. 2013; reviewed in Engler et al. 2018). These cells are characterized by a high proliferation rate, transforming to intermediate progenitors that amplify the progenitor pool and neuronal progeny, a process influenced by Notch signaling (Noctor et al. 2004; reviewed in Engler et al. 2018). Under the control of a tightly regulated network of various pathways, including

Notch, these committed neural progenitors migrate out of the VZ and colonize the subventricular zone where they divide and generate neuroblasts, thus amplifying the progenitor pool and consequently the number of neuronal progeny (reviewed in Engler et al. 2018). The neuroblasts migrate radially through the cortex forming layers in an inside-out fashion according to their age (reviewed in Engler et al. 2018). While neuroblasts of projection neurons migrate along the radial glial fibers of the radial glia to the superficial layers in the pallium, the interneurons are generated in the ganglionic eminences of the subpallium and undergo a long-range tangential migration to reach the cortex (Marin 2013, reviewed in Engler et al. 2018). During the neurogenic phase of brain development (embryonic neurogenesis), and under the control of a tightly regulated network of various pathways, including Notch, NSCs have to divide to generate differentiated progeny but also maintain the stem cell pool (reviewed in Engler et al. 2018). Hence, following cell division one of the daughter cells must remain as a NSC. Some of these retained NSCs are maintained from the embryo even into the adult brain where they later can function as a source for adult-born neurons (Fuentealba et al. 2015; Furutachi et al. 2015; reviewed in Engler et al. 2018). Adult NSCs are found in two distinct niches, the lateral wall (LW) of the subventricular zone (SVZ) and the subgranular zone (SGZ) of the dentate gyrus (DG) (reviewed in Engler et al. 2018). Adult NSCs generate differentiated neurons through intermediate progenitors/transient amplifying progenitors that rapidly divide and consecutively give rise to neuroblasts and neurons (reviewed in Engler et al. 2018). Under the control of a tightly regulated network of various pathways, including Notch, neuroblasts of the LW migrate along the rostral migratory stream (RMS) into the olfactory bulb (OB) where they differentiate and integrate into local circuits, while the adult-born neurons in the SGZ integrate locally (reviewed in Engler et al. 2018).

It was reported that the tightly regulated and balanced interplay between many different signaling pathways, including sonic hedgehog (Shh), wingless (Wnt), retinoic acid (RA), fibroblast growth factor (FGF), and bone morphogenetic protein (BMP), induces and regulates the formation of the neural tube by NEPs (Greig et al. 2013; Franco and Muller 2013; Lupo et al. 2006; reviewed in Engler et al. 2018) and its regionalization in several well-defined structural domains including four important segments, namely, the forebrain, the midbrain, the hindbrain, and the spinal cord. Notably, the forebrain includes two important cortical structures – the neocortex and the hippocampus – that are both generated embryonically and in the early postnatal period (reviewed in Engler et al. 2018). It has been shown that the generation of the neocortex, which involves Notch signaling, begins by E11.5 and that at least its neuronal architecture is finished by birth, whereas the generation of the hippocampus begins by E17.5 and is in mice anatomically complete around postnatal day 14 (P14) (Nicola et al. 2015; Rolando and Taylor 2014; reviewed in Engler et al. 2018). It has been demonstrated that these two anatomical regions also contain the NSC niches and supply the known neurogenic regions of the adult brain with stem cells. In mice, the neuroepithelium represents at E9 a pseudostratified single layer of NEPs (reviewed in Engler et al. 2018).

When the generation of neurons from the neuroepithelium, which involves Notch signaling, begins, the NEPs transform into radial glia cells (RGCs), which generate the VZ and function during embryonic development as NSCs (Noctor et al. 2004, reviewed in Engler et al. 2018). These populations of precursor and progenitor cells show characteristic features of the embryonic brain (reviewed in Engler et al. 2018). While the soma of RGCs remains in the VZ, these cells span with their radial process the thickness of the cortex from the apical to the basal surface (Gotz and Huttner 2005, reviewed in Engler et al. 2018). RGCs are characterized by a polarized structure spanning the thickness of the neural tube with an apical process anchored at the lumen of the tube and a long basolateral process to the forming surface of the brain (reviewed in Engler et al. 2018). It has been shown that intermediate progenitor cells (IPCs) are the progeny of RGCs (reviewed in Engler et al. 2018). The SVZ harbors this characteristic cell population, which is not connected to either surface of the neural tube (reviewed in Engler et al. 2018). While RGCs represent the NSCs of the mammalian brain, it has been shown that IPCs are short-lived intermediate cells (reviewed in Engler et al. 2018). The generation of the neocortex, which involves Notch signaling, can be separated into an initial expansion period followed during mid-late embryogenesis (E9-E18) by a neurogenic period and then by a gliogenic period (reviewed in Engler et al. 2018). Under the control of a tightly regulated network of various pathways, including Notch, NSCs undergo, during the expansion period, symmetric cell divisions, while during the neurogenic period of embryonic development, NSCs primarily divide asymmetrically, thereby generating one daughter cell that remains as a stem cell and another daughter cell that progresses to an IPC that is committed to differentiation (Noctor et al. 2007; Noctor et al. 2004, reviewed in Engler et al. 2018). Notably, before E15.5, direct neurogenesis of NSCs has been reported, generating neurons independent from the presence of an IPC (Telley et al. 2016). The excitatory neurons of the neocortex are produced in a sequential manner by the NSCs to generate in an inside-out fashion the characteristic six individual layers of the isocortex (Franco and Muller 2013; Guo et al. 2013; reviewed in Engler et al. 2018). During the later steps of the neurogenic period, the NSCs switch in their fate and begin to form glial cells (reviewed in Engler et al. 2018). Interestingly, a part of the NSC population is separated and reserved during embryonic development of the neocortex as a supply for the adult NSC pools (reviewed in Engler et al. 2018). It has been shown that this characteristic population of cells exit cell cycle during the embryogenesis and remain relatively inactive until postnatal and adult periods (Fuentealba et al. 2015; Liu et al. 2011; Greig et al. 2013; Furutachi et al. 2015; reviewed in Engler et al. 2018).

Embryonic Neural Stem Cells and Notch Signaling

It has now been convincingly demonstrated that Notch signaling represents one of the key control mechanisms that govern maintenance of NSCs (reviewed in Engler et al. 2018). In the embryonic brain, NSCs express Notch receptors, and via mechanisms that involve active sig-(NICD)-Rbpj-Maml, expression of the basic helix-loop-helix factors HES1 and HES5 is induced (Honjo 1996; Artavanis-Tsakonas et al. 1999; Ohtsuka et al. 1999; Kageyama et al. 2007; reviewed in Engler et al. 2018) (Table 2.1). One important function of HES1 and HES5 is the suppression of the expression of various proneural genes (*Ascl1*, *Atoh1*, *Neurog1*, and *Neurog2*), which in turn blocks NSC differentiation and neuron production (Lutolf et al. 2002; Hatakeyama et al. 2004; Hatakeyama et al. 2001; reviewed in Engler et al. 2018) (Table 2.1). It has now been demonstrated that, as in the formation of the somites, Notch signaling and *Hes1* expression fluctuate in neural stem/progenitor cells of the developing embryo (Masamizu et al. 2006; Shimojo et al. 2008, reviewed in Engler et al. 2018), a process for which the cyclical synthesis and degradation of Hes1 mRNA and protein are of high importance. Moreover, it was shown that HES1 binds its own promoter and thereby suppresses its own expression, counteracting activation of Notch (reviewed in Engler et al. 2018). In neural stem/progenitor cells, the period of *Hes1* pulsatory expression is around 2–3 h, and this periodicity of fluctuation projects onto the expression of the proneural genes and Delta-like 1 (*Dll1*), which represents a repressed target gene of Notch (via Hes-mediated suppression (Hirata et al. 2002, reviewed in Engler et al. 2018)). However, in differentiating neurons *Hes* expression is suppressed, resulting in sustained expression of *Ngn2* and *Dll1* (Shimojo et al. 2008) (Table 2.1). Notably, the Hes-mediated Notch signaling feedback mechanism governs neural fate separation into stem/progenitor and neurons (Shimojo et al. 2011) (Table 2.1). Interestingly, deletion of either *Hes1* or *Hes5* has during embryonic development no obvious effects on neural development and NSCs activity (reviewed in Engler et al. 2018). However, it has been shown that the simultaneous deletion of *Hes1* and *Hes5* results in distinct phenotypes characterized by severe pathologies, causing in the embryo disorganization of the neural tube, premature neuronal differentiation, and loss of radial glia (Hatakeyama et al. 2004; reviewed in Engler et al. 2018). Moreover, *Hes1* has previously been identified to represent an important target of BMP-mediated and other signaling pathways (Kageyama et al. 2007). In summary, a broad body of scientific evidence, including the findings outlined above, now convincingly demonstrates the important function of the Notch pathway for the maintenance of NSCs during the development of the brain (reviewed in Engler et al. 2018).

Unraveling Hidden Secrets: Notch Signaling in the Embryonic Development of Skin and Hair Follicles

A large body of evidence has convincingly shown the importance of Notch signaling for the embryonic development of all anatomical structures of the skin, including the epidermal compartment and appendages such as hair follicles (HF). It was demonstrated that, in response to external cues, embryonic skin cells have to make a cell fate decision whether or not to differentiate and generate stratified epidermis,

Table 2.1 Notch signaling in embryonic development: lessons learned from in vitro investigations and animal studies

Organ	Molecular target	Intervention and biological effect	References
Central nervous system (CNS)/brain	Hes-mediated Notch signaling feedback mechanism	Governs neural fate separation into stem/progenitor cells and neurons	Shimojo et al. (2011) Reviewed in Engler et al. (2018)
	Notch receptors	Expressed in embryonic brain, induce via mechanisms that involve active sig-(NICD)-Rbpj-Maml expression of HES1 and HES5 who suppress expression of proneural genes (*Ascl1, Atoh1, Neurog1*, and *Neurog2*)	Reviewed in Engler et al. (2018)
	Notch1	KO: in embryonic neurogenesis lethal (E9.5)	Conlon et al. (1995)
		GOF: in embryonic neurogenesis glial, instead of neuronal fate	Reviewed in Engler et al. (2018)
		cKO: in embryonic neurogenesis precocious cell cycle exit, neurogenesis increased cKO: in adult neurogenesis block of NSCs self-renewal	Gaiano et al. (2000) Lutolf et al. (2002) Mason et al. (2006) Reviewed in Engler et al. (2018)
	Notch2	KO: in embryonic neurogenesis lethal (E9.5)	Hamada et al. (1999) Reviewed in Engler et al. (2018)
	Rbpj	KO: in embryonic neurogenesis lethal (E10.5), delayed CNS development KO: in adult neurogenesis depletion and exhaustion of NSCs	de la Pompa et al. (1997) Reviewed in Engler et al. (2018)
	Hes1	In embryonic neurogenesis: expression suppressed in differentiating neurons, resulting in sustained expression of *Ngn2* and *Dll1*	Shimojo et al. (2008) Reviewed in Engler et al. (2018)
		KO: in embryonic neurogenesis redundancy by Hes5	Hatakeyama et al. (2004) Reviewed in Engler et al. (2018)
		Compound KO Hes1 and Hes5: in embryonic neurogenesis loss of NSCs	Reviewed in Engler et al. (2018)
	Hes5	KO: in embryonic neurogenesis redundancy by Hes1 In adult neurogenesis expressed in NSCs and astrocytes	Reviewed in Engler et al. (2018)
	Jag	K0: in embryonic neurogenesis lethal (E10.5)	Xue et al. (1999) Reviewed in Engler et al. (2018)
		cKO: in embryonic neurogenesis defects in migration, differentiation, survival cKO: in adult neurogenesis inhibition of NSCs self-renewal	Weller et al. (2006) Reviewed in Engler et al. (2018)

(continued)

Table 2.1 (continued)

Organ	Molecular target	Intervention and biological effect	References
	Dll1	In embryonic neurogenesis expressed till E13.5	Stump et al. (2002) Reviewed in Engler et al. (2018)
		cKO: in adult neurogenesis loss of quiescent NSCs	
	Dll4	In embryonic neurogenesis weak expression	Stump et al. (2002) Reviewed in Engler et al. (2018)
		GOF: in adult neurogenesis stimulated NSC proliferation and survival	
Kidney	Notch2ICD	Overexpression increases Wnt4 expression at E11.5, leading to premature tubule differentiation and depletion of nephron progenitors by E14.5 and subsequently to the formation of multiple cysts and deterioration of the kidney	Fujimura et al. (2010) Reviewed in Mašek and Andersson (2017)
Liver	Jag1	Expression arises around E12.5 in the portal vein mesenchyme and is associated with the onset of embryonic bile duct development	Hofmann et al. (2010) Zong et al. (2009) Reviewed in Mašek and Andersson (2017)
Pancreas	Jag1	Conditional deletion in pancreatic epithelial cells results in abnormal ductal formation, fibrosis, and chronic pancreatitis Governs embryonic pancreas development during embryonic stages via inhibition of Dll1-Notch signaling and during postnatal stages via stimulation of Notch signaling	Golson et al. (2009b) Reviewed in Mašek and Andersson (2017)
	Notch1 and Notch2	Conditional compound deletion in pancreatic epithelial cells results only in weak effects on pancreatic epithelial cell proliferation	Nakhai et al. (2008) Reviewed in Mašek and Andersson (2017)
Heart	Notch1	Ablation has negative effect on the formation of coronary vasculature in the compact myocardium Increased expression of Notch1ICD abrogates subepicardial ECM, reduces thickness of compact myocardium, and disturbs integrity of the myocardium	MacGrogan et al. (2016) Del Monte et al. (2011) Reviewed in Mašek and Andersson (2017)
	Jag1	In the endocardium, ablation results in outflow tract defects, aortic valve hyperplasticity, tetralogy of Fallot, and valve calcification Governs, in combination with Jag2, the maturation and compaction of the chamber myocardium	
	Jag2	Governs, in combination with Jag1, the maturation and compaction of the chamber myocardium	

(continued)

Table 2.1 (continued)

Organ	Molecular target	Intervention and biological effect	References
Skeleton	Jag1	Ablation results in progressive bone loss in adult mice	Canalis et al. (2016) Zanotti and Canalis (2013)
	Notch2	Ablation results in progressive bone loss in adult mice Mice harboring a gain-of-function mutation (Q2319X) exhibit enhanced osteoclastogenesis, leading to cancellous and bone osteopenia and increased bone resorption	Nobta et al. (2005) Hilton et al. (2008) Youngstrom et al. (2016) Zanotti and Canalis (2013) Reviewed in Mašek and Andersson (2017)
Eye	Notch2	Deletion (via lens-cre) disrupts lens differentiation	Le et al. (2012) Reviewed in Mašek and Andersson (2017)
Ear	Jag1	Mutant mouse strains (named Slalom, Headturner, Ozzy, and Nodder) show impaired balance and deafness	Tsai et al. (2001) Hansson et al. (2010) Kiernan et al. (2001) Vrijens et al. (2006) Reviewed in Mašek and Andersson (2017)
Skin	Notch receptors and corresponding ligands	Around E13.5, cell fate decisions of epidermal keratinocytes whether or not to transit from basal to suprabasal cell layers begin and are associated with stratification of the epidermis and activation of Notch receptors by corresponding ligands, which is on the molecular level mediated by enzymatic cleavage of NICD and its translocation to the nucleus, where it associates in suprabasal keratinocytes with DNA-binding protein RBP-j to regulate downstream target genes	Blanpain et al. (2006)

Abbreviations: *cKO* conditional knockout, *Dll* Delta-like, *E* embryonic day, *GOF* gain of function, *Hes* hairy and enhancer of split, *jag* jagged, *KO* knockout, *NICD* Notch intracellular domain, *NSC* neural stem cell

or to invaginate and initiate morphogenesis of HF (Fuchs 2007). In the skin, the epidermis is maintained throughout life through the proliferation of stem cells and differentiation of their progeny. The innermost (basal) layer of the epidermis consists of proliferative progenitor cells which give rise to multiple differentiating layers, a stratified epithelium providing a barrier that keeps the inside of the body moist and protects the body from outside assaults by physical, environmental, and biological factors (Massi and Panelos 2012). Data from ongoing studies indicate that Notch signaling orchestrates the process of epidermal differentiation and proliferation through the sequential activity of different Notch ligands, receptors, and downstream pathways. Investigations using transgenic mice have demonstrated that in contrast to embryonic development of the HF that can be achieved without Notch, its postnatal development requires an intact Notch signaling in two important compartments of the hair, the bulb and the outer root sheath (reviewed in Aubin-Houzelstein 2012, reviewed in Massi and Panelos 2012). In the hair bulb, Notch governs cell differentiation, ensuring the proper development of every layer of both the hair shaft and the inner root sheath (reviewed in Aubin-Houzelstein 2012, reviewed in Massi and Panelos 2012). Among the many roles played by Notch in the skin and HF, it has to be highlighted that in the bulge, Notch controls a cell fate switch in hair follicle stem cells or their progenitors, preventing them from adopting an epidermal fate (reviewed in Aubin-Houzelstein 2012). Notch function in the skin and HF is both cell autonomous and cell nonautonomous and involves intercellular communication between adjacent cell layers (reviewed in Aubin-Houzelstein 2012, reviewed in Massi and Panelos 2012). This tightly regulated process depends on Wnt-mediated signals from adjacent epidermal cells and suppressing bone morphogenetic protein (BMP)-mediated signals from underlying mesenchymal condensates, which converge to activate sonic hedgehog (Shh) in the developing hair bud. Loss of Shh signaling widely disturbs this highly regulated epithelial-mesenchymal crosstalk, impairing HF downgrowth and maturation in the embryo and distorting homeostasis throughout postnatal skin epithelium (Chiang et al. 1999; Gritli-Linde et al. 2007; Oro and Higgins 2003). Notably, it was shown that epidermal morphogenesis not only precedes but also may be observed independent of Hh signaling (Oro and Higgins 2003).

It has been demonstrated that the cell fate decisions of epidermal keratinocytes whether or not to transit from basal to suprabasal cell layers begin around embryonic day 13.5 (E13.5). At this time, the activation of Notch receptors by their corresponding ligands is associated with the stratification of the epidermis (Blanpain et al. 2006). On the molecular level, this process is mediated by enzymatic cleavage of Notch intracellular domain (NICD) and its translocation to the nucleus, where it associates in keratinocytes of suprabasal cell layers with DNA-binding protein RBP-j to regulate downstream target genes (Kopan and Ilagan 2009; Lowell et al. 2000; Moriyama et al. 2008; Okuyama et al. 2004; Wang et al. 2008).

During the embryonic development of the epidermis, Notch signaling triggers a terminal differentiation program that culminates in skin barrier formation (Nguyen et al. 2006). Investigations using transgenic mice carrying loss-of-function mutations have shown that NICD-RBP-j activation is essential for the early transition of basal progenitors to committed, suprabasal "spinous" cells, a switch that is characterized by downregulation of cytokeratins K5/K14 and induction of cytokeratins K1/K10 and by dramatic downstream architectural changes in cytoskeletal and intercellular adhesion (Blanpain et al. 2006). However, little is known about the events residing upstream of Notch-NICD-RBP-j that induce its activation.

The importance of Notch signaling for skin embryogenesis is underlined by Adams-Oliver syndrome, a rare genetic disorder that has been linked to mutations in several different genes, including *DLL4* (OMIM 605185; cytogenetic location: 15q15.1) and *NOTCH*1 (OMIM 190198; cytogenetic location: 9q34.3), as well as in RBPJ (OMIM 147183; cytogenetic location: 4q15.2), *EOGT* (OMIM 614789; cytogenetic location: 3p14.1), *ARHGAP*31 (OMIM 610911; cytogenetic location:

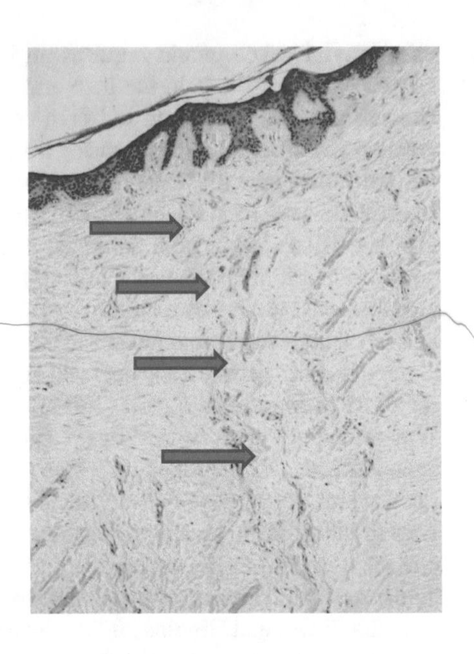

Fig. 2.1 Skin histology of aplasia cutis congenita, a clinical hallmark of Adams-Oliver syndrome. Please note dermal atrophy with lack of elastic fibers and rarefication of epidermal appendages and other skin structures

3q13.2-3q13.33), and *DOCK6* (OMIM 614194; cytogenetic location: 19p13.2) (reviewed in Mašek and Andersson 2017; Meester et al. 2019). Adams-Oliver syndrome is diagnosed based on the presence of aplasia cutis congenita and several other clinical hallmarks, namely, terminal transverse limb malformations, and a partial absence of skull bones (reviewed in Mašek and Andersson 2017; Meester et al. 2019; Zanotti and Canalis 2016). Typically, aplasia cutis congenita is found in the skull region; however, other body parts, including the abdomen, may also be affected (reviewed in Meester et al. 2019; Zanotti and Canalis 2016). The severity and symptoms of aplasia cutis congenita may greatly vary (reviewed in Meester et al. 2019; Zanotti and Canalis 2016). At birth, the affected skin region typically presents as healed but scarred skin, and skin histology (Fig. 2.1) shows characteristic findings that may include absent epidermis, dermal atrophy, and a lack of elastic fibers and other skin structures (reviewed in Meester et al. 2019; Zanotti and Canalis 2016). However, symptoms may range from a localized region with complete absence of skin to patches of skin that lack hair (reviewed in Meester et al. 2019; Zanotti and Canalis 2016).

From the Bench to the Patient: The Impact of the Notch Pathway for Embryonic Kidney Development, Notch2, Jag1, and Beyond

Embryonic development of the kidneys is tightly controlled by Notch signaling. Consequently, many patients with Alagille syndrome present with serious kidney problems (Kamath et al. 2013, reviewed in Mašek and Andersson 2017), although this is not a diagnostic criterion. During embryonic kidney development, expression of Notch2 arises first in the branched ureteric bud and the surrounding cap mesenchyme, while Jag1 is first expressed in epithelial vesicles (the aggregates derived from cap mesenchyme via mesenchymal-to-epithelial transition – MET), together with its corresponding receptors Notch2 and Notch1. These vesicles transform through the stages of comma-shaped bodies and S-shaped bodies into fully developed nephrons in which Jag1 is expressed in the glomerular endothelium, and both Notch1 and Notch2 are expressed in glomerular epithelial cells (reviewed in Kamath et al. 2013; Kopan et al. 2014, reviewed in Mašek and Andersson 2017). In line with these expression patterns, it has been demonstrated that mice haplo-insufficient for *Notch2* and lacking one allele of *Jag1* show defective glomerulogenesis (reviewed in Mašek and Andersson 2017, McCright et al. 2001), while the cap mesenchyme-specific depletion of the corresponding receptor Notch2, but not of the receptor Notch1, leads postnatal to early lethality, which is caused by the blockade of the development of podocytes and proximal tubules prior to S-shaped body formation (Cheng et al. 2007, reviewed in Mašek and Andersson 2017). Notably, receptors Notch1 and 2 can both be activated by their corresponding ligands Jag1 or Dll1 (Liu et al. 2013, reviewed in Mašek and Andersson 2017). It has been speculated that this unequal requirement for signaling mediated by Notch receptors 1 and 2 during renal embryonic development may likely be caused by differences in their extracellular domains and/or by interaction with the Lfng. Indeed, Lfng enhances Notch2-mediated signaling to a greater extent as compared to Notch1-induced signaling, and it has been speculated that it therefore may be an important factor required to gain the threshold needed for stimulation of proximal structure formation (Liu et al. 2013, reviewed in Mašek and Andersson 2017). However, this hypothesis remains to be confirmed in genetic experiments. One should also keep in mind that, although Notch signaling-independent MET has been reported (Cheng et al. 2007; Chung et al. 2016, reviewed in Mašek and Andersson 2017), the Notch pathway can compensate for Wnt/β-catenin signaling during MET (Boyle et al. 2011, reviewed in Mašek and Andersson 2017), and its activity in medial and proximal segments, which is stimulated by BMP signaling, is mutually exclusive, with high activity of the Wnt/β-catenin pathway (Lindström et al. 2015, reviewed in Mašek and Andersson 2017). While the investigations described above underline key roles for receptor Notch2 and its corresponding ligand Jag1 in embryonic kidney development, it is not known in humans how individual mutations associated with syndromes linked to Notch signaling cause kidney defects. In this context, it has to be noted that no kidney phenotype has been reported in the Hajdu-Cheney syndrome mouse model harboring the Notch2^{Q2319X} mutation (Canalis et al. 2016, reviewed in Mašek and Andersson 2017). Moreover, several investigations convincingly demonstrate that high levels of Notch signaling negatively affect embryonic kidney development. In line with this assumption, it was shown in the metanephric mesenchyme that constitutively active Notch1 intracellular domain (ICD) (Cheng et al. 2007, reviewed in Mašek and Andersson 2017) or Notch2ICD (Six2-GFP: Cre) (Fujimura et al. 2010, reviewed in Mašek and Andersson 2017) promotes pathological embryonic kidney development. In contrast to overexpression of Notch1ICD, which promotes single ureteric bud formation and is associated with greatly increased proximal tubule transformation into podocytes and distal tubules (at the expense of mesenchymal progenitor differentiation) (Cheng et al. 2007, reviewed in Mašek and Andersson 2017), it was shown that overabundance of Notch2ICD stimulates *Wnt4* expression at E11.5 (Table 2.1), leading to premature tubule differentiation and depletion of nephron progenitors by E14.5, a process that is followed by formation of multiple cysts and general deterioration of the kidney (Fujimura et al. 2010, reviewed in Mašek and Andersson 2017).

Notch Receptor 2 and Its Corresponding Ligand Jag1 Are Key Players for Embryonic Liver Development

In mammalians and humans, embryonic liver development is a tightly regulated, complex process that is controlled by a network of many signaling pathways, including the Notch pathway, which is closely linked with the development of bile ducts (reviewed in Gordillo et al. 2015, reviewed in Mašek and Andersson 2017). Notably, it has been shown in mice that when expression of Jag1 arises in the portal vein mesenchyme (PVM) around E12.5, this process is associated with the onset of embryonic bile duct development (Hofmann et al. 2010; Zong et al. 2009, reviewed in Mašek and Andersson 2017) (Table 2.1). It has been demonstrated that Jag1 stimulates the expression of Hes1, Hnf1β, and Sox9 via binding to its corresponding receptor Notch2 in adjacent biliary epithelial cells. The upregulation of these genes then further controls the morphogenesis of the intrahepatic bile duct (Antoniou et al. 2009; Geisler et al. 2008; Kodama et al. 2004; Zong et al. 2009, reviewed in Mašek and Andersson 2017). Notably, it has been demonstrated previously that the presence of Jag1 is not required in biliary epithelial cells (Loomes et al. 2007, reviewed in Mašek and Andersson 2017) and in the portal vein endothelium (PVE) (Hofmann et al. 2010, reviewed in Mašek and Andersson 2017), but specifically in the portal vein mesenchyme (PVM). However, presence of Jag1 has also been shown in these anatomical structures, namely, biliary epithelial cells and PVE. Interestingly, it has been found that hypomorphic mice heterozygous for *Jag1* and *Notch2* mimic several characteristic features of Alagille syndrome, including jaundice, growth retardation, disrupted differentiation of intrahepatic bile ducts, and heart, eye, and kidney defects (McCright et al. 2002, reviewed in Mašek and Andersson 2017). Moreover, it was shown that backcrossing of *Jag1*$^{+/-}$ into a C57BL/6J background leads to defects similar to those observed in *Jag1/Notch2* double heterozygotes (reviewed in Mašek and Andersson 2017, Thakurdas et al. 2016), indicating that the biliary phenotype highly depends on the genetic background, similar to other Jag1 phenotypes (Kiernan et al. 2007, reviewed in Mašek and Andersson 2017).

In that study, it was also demonstrated that stability of Jag1 is suppressed by O-glucosyltransferase 1 (POGLUT1, also named *Rumi*). Moreover, this suppression of Rumi rescues in *Jag1*$^{+/-}$/*Rumi*$^{+/-}$ animals the biliary phenotype of *Jag1*-deficient animals. Deficiency of *Notch2* results perinatally in agenesis of the bile duct and after weaning to secondary bile duct formation (Falix et al. 2014, reviewed in Mašek and Andersson 2017), a process that presumably does not depend on Notch signaling (reviewed in Mašek and Andersson 2017, Walter et al. 2014). Interestingly, in Alagille syndrome, a similar recovery of the liver phenotype with age was demonstrated (reviewed in Mašek and Andersson 2017, Riely et al. 1979). However, it is unknown which *JAG1* or *NOTCH2* genotypes, if any, are involved in this recovery.

The Role of Notch Signaling for Embryonic Development of the Pancreas

Interestingly, pancreatitis and other disorders characterized by pathological pancreatic function have been associated with Alagille syndrome (reviewed in Mašek and Andersson 2017, Rovner et al. 2002, Devriendt et al. 1996). However, recent investigations using different methodological approaches demonstrated alterations in pancreatic function in only two out of forty-two patients with Alagille syndrome (Kamath et al. 2012, reviewed in Mašek and Andersson 2017). Notably, it has been reported that *Notch2* and *Jag1* are of high importance for murine embryonic pancreas development and that their impaired function may in humans, at least in some patients, be responsible for pancreas defects. Notch signaling governs both the primary (Ahnfelt-Rønne et al. 2012; Jensen et al. 2000, reviewed in Mašek and Andersson 2017) and secondary (reviewed in Mašek and Andersson 2017, Murtaugh et al. 2003; Shih et al. 2012) waves of pancreatic progenitor differentiation, which take place at E8.5-E12.0 and at E13.0-E16.0 of mouse gestation, respectively. These events lead to full set of endocrine (α-, β-, δ-, ε-, and PP-cells), acinar, and duct cells (Li et al. 2016; reviewed in Afelik and Jensen 2013, reviewed in Mašek and Andersson 2017). Jag1 governs embryonic pancreas development via inhibition of Dll1-Notch signaling during embryonic stages and via stimulation of Notch signaling during postnatal stages (Golson et al. 2009a, reviewed in Mašek and Andersson 2017) (Table 2.1). Conditional deletion of pancreatic epithelial *Jag1* (using Pdx1-Cre) results in abnormal ductal formation, fibrosis, and chronic pancreatitis (Golson et al. 2009b, reviewed in Mašek and Andersson 2017) (Table 2.1). Remarkably, conditional compound deletion of *Notch1* and *Notch2* results only in weak effects on pancreatic epithelial cell proliferation (Nakhai et al. 2008, reviewed in Mašek and Andersson 2017). It has been speculated that this observation can be explained by a rescue of the phenotype by Notch3 (Apelqvist et al. 1999, reviewed in Mašek and Andersson 2017).

Notch2 and Jag1 Function in Embryonic Heart Development

It has been shown that heart embryogenesis depends on a tightly regulated complex interplay of morphogenesis events, including tube formation and looping (reviewed in Sedmera 2011, reviewed in Mašek and Andersson 2017). On the cellular level, these events are associated with multiple cell fate decisions in many different cell types involved, including proliferation, differentiation, and migration. Jag1 and its corresponding receptor Notch2 are both expressed from early stages of the formation of the heart and – together with other components of the Notch signaling pathway (reviewed in D'Amato et al. 2016a; reviewed in Luxán et al. 2016, reviewed in Mašek and Andersson 2017) – govern different key steps of embryonic cardiac development. Although it is still not known how distinct *JAG1* mutations, which have variable effects on JAG1 trafficking and activity, can be linked to the range of cardiac defects observed in individuals with Alagille syndrome (Bauer et al. 2010, reviewed in Mašek and Andersson 2017), it has been demonstrated that balanced activities of Jag1 and Notch2 are required for the development of several compartments of the heart.

Notch signaling governs many steps of embryonic mammalian heart development, being present and exerting important activities in various tissue types and compartments. It has been reported that loss-of-function mutations in *NOTCH1* (Garg et al. 2005; Theodoris et al. 2015, reviewed in Mašek and Andersson 2017) and the E3 ubiquitin ligase mind bomb1 (*MIB1*) (Luxán et al. 2013; reviewed in Mašek and Andersson 2017) are of importance for calcific aortic valve disease (CAVD) and left ventricular noncompaction (LVNC) congenital cardiovascular diseases, respectively. Dll4 is an important stimulator of its corresponding receptor Notch1 that governs endothelial-to-mesenchymal transformation (EndoMT) (MacGrogan et al. 2016; reviewed in Mašek and Andersson 2017). Because Dll4 expression in the endocardium is reduced with the progression of endocardial cushion formation around E10, Jag1/Notch1 signaling-mediated expression of heparin-binding epidermal growth factor (EGF)-like growth factor (Hbegf) becomes important to inhibit the BMP-mediated proliferation of cardiac valve mesenchyme (MacGrogan et al. 2016; reviewed in Mašek and Andersson 2017). Jag1, in combination with Jag2, also governs the maturation and compaction of the myocardium of the ventricular chamber. At first, Jag1/Jag2-induced activation of their corresponding receptor Notch1 is inhibited by Dll4 and Mfng in the endocardium, but after E11, expression of Dll4 and Mfng is reduced and Jag1/2 can stimulate signaling mediated by its corresponding receptor Notch1, promoting proliferation and compaction of the chamber myocardium. The Notch pathway also has a role in the epicardium, which is an important source of cells for coronary vessel formation (reviewed in Perez-Pomares and de la Pompa 2011; reviewed in Mašek and Andersson 2017). It has been shown that, as a prerequisite for correct heart development, Notch signaling in/from the epicardium needs to be in an equilibrium (Grieskamp et al. 2011; reviewed in Mašek and Andersson 2017). In line with this finding, the ablation of Notch1 has a negative impact on the formation of coronary vasculature in the compact myocardium, while increased expression of Notch1ICD abrogates subepicardial ECM, reduces the thickness of compact myocardium, and disturbs the integrity of the epicardium (Del Monte et al. 2011, reviewed in Mašek and Andersson 2017). However, the precise functions of Jag1 and its corresponding receptor Notch2, which are also present in the epicardium, and Notch3 and its corresponding ligand Dll4, which are present in epicardium-derived vSMCs, need to be further investigated.

Moreover, ablation of *Jag1* expression in the endocardium results in outflow tract (OFT) defects, aortic valve hyperplasticity, tetralogy of Fallot, and valve calcification (Table 2.1), summarizing the spectrum of cardiac pathologies often found in Alagille syndrome (Hofmann et al. 2012; MacGrogan et al. 2016; reviewed in Mašek and Andersson 2017). These characteristic phenotypes are, at least in part, connected to cardiac neural crest (CNC) cells, a cell population with high migratory activity that develops from the neural plate border (Jiang et al. 2000; reviewed in Mašek and Andersson 2017). Investigations using CNC-specific deletion (using Pax3-Cre) of either *Jag1* (Manderfield et al. 2012; reviewed in Mašek and Andersson 2017) or *Notch2* (Varadkar et al. 2008; reviewed in Mašek and Andersson 2017) demonstrated that they are not a prerequisite for CNC migration, but that Jag1 is an important stimulator of CNC-derived vSMC differentiation. Notch2-mediated signaling, meanwhile, sustains vSMC proliferation around the aortic arch arteries and OFT (Varadkar et al. 2008; reviewed in Mašek and Andersson 2017). Moreover, impaired Jag1 and Notch2 signaling also causes ventricular septation defects, aortic arch patterning defects, and pulmonary artery stenosis, all of which are conditions found in patients with Alagille syndrome (Manderfield et al. 2012; Varadkar et al. 2008; reviewed in Mašek and Andersson 2017). Another possible reason for the congenital heart disease present in individuals with Hajdu-Cheney syndrome (Crifasi et al. 1997; reviewed in Mašek and Andersson 2017) is delivered by the role of Notch2 in the formation of trabecular myocardium: under physiological conditions, Notch2 activity must be inhibited by Numb and Numbl to balance the formation of compact versus trabecular myocardium, and its increased expression results in hypertrabeculation, noncompaction, and septation defects (Yang et al. 2012; reviewed in Mašek and Andersson 2017). Future investigations are needed to determine how *Notch2* exerts these functions, keeping in mind that *Notch2* mRNA expression is absent in the developing myocardium (D'Amato et al. 2016b; reviewed in Mašek and Andersson 2017).

Notch2 and Jag1 Function During Skeletal Development

It has been convincingly demonstrated that Notch signaling is of high importance for the developing skeleton and, consequently, skeletal defects are present in many congenital Notch disorders, including Alagille and Hajdu-Cheney syndromes and spondylocostal dysostosis (reviewed by Zanotti and Canalis 2016; reviewed in Mašek and Andersson 2017). However, it has to be noted that the systemic deletion of *Jag1* or *Notch2* does not cause any characteristic somite-related phenotype that would indicate a contribution of these mediators of Notch signaling to the early events of bone formation (Hamada et al. 1999; Xue et al. 1999; reviewed in Mašek and Andersson 2017). It has been shown that both Jag1 and Notch2 negatively affect the differentiation of mesenchymal progenitors into osteoblasts in the skeletogenic mesenchyme, an observation that was reported in vitro and in adolescent mice. Moreover, it was reported that ablation of Jag1 and Notch2 results in progressive bone loss in adult mice (Hilton et al. 2008; Nobta et al. 2005; Youngstrom et al. 2016;

reviewed in Mašek and Andersson 2017) (Table 2.1). Notably, deletion of *Jag1* leads in mesenchymal progenitors to extension of the cortical bone, while reducing trabecular bone mass, indicating contrary effects of Jag1-mediated Notch signaling on cortical as compared with trabecular osteoblasts (Youngstrom et al. 2016; reviewed in Mašek and Andersson 2017). This disequilibrium then causes spine defects and the formation of butterfly vertebrae, a typical feature of Alagille syndrome (Emerick et al. 1999; Youngstrom et al. 2016; reviewed in Mašek and Andersson 2017). Additionally, both clinical and genome-wide association studies suggest an association between mutations in *JAG1* and reduced bone mineral density as well as osteoporotic fractures (Bales et al. 2010; Kung et al. 2010; reviewed in Mašek and Andersson 2017). The formation of craniofacial bone, which develops from intramembranous ossification of neural crest (NC)-derived mesenchyme, also depends on Jag1: its ablation in NC cells disturbs mesenchymal differentiation and results in abrogated mineralization and deformities of the craniofacial skeleton, another feature found both in patients with Alagille and Hajdu-Cheney syndromes (Hill et al. 2014; Humphreys et al. 2012; reviewed in Mašek and Andersson 2017).

Recently, gain-of-function *Notch2* mice harboring a Q2319X mutation were demonstrated to exhibit enhanced osteoclastogenesis, leading to cancellous and cortical bone osteopenia and increased bone resorption (Canalis et al. 2016; reviewed in Mašek and Andersson 2017). This characteristic is strikingly different from the phenotypes observed in odontoblast- and osteocyte-specific Notch1ICD gain-of-function mice (Canalis et al. 2013; reviewed in Mašek and Andersson 2017). This variation might be the result of differences between constitutive and Cre-dependent methodological approaches, different levels of Notch activation, or other unknown factors extrinsic to skeletogenic mesenchyme that cause the Hajdu-Cheney syndrome phenotype.

The Impact of the Notch Pathway for Angiogenesis

It has convincingly been shown that components of the Notch signaling pathway govern various important aspects of vascular development, from vascular growth and endothelial tip and stalk cell selection to vSMC development. Because of resulting defects in angiogenesis of the embryonic and yolk sac vasculature, the systemic knockout of *Jag1* is embryonic lethal in mice at ~E11.5 (Kiernan et al. 2007; Xue et al. 1999; reviewed in Mašek and Andersson 2017). A similar picture is found in homozygous *Notch2* knockout mice, which are characterized by widespread apoptosis and die at ~E10.5 (Hamada et al. 1999; McCright et al. 2006; reviewed in Mašek and Andersson 2017). The endothelial-specific ablation (via Tie1- or Tie2-Cre) of *Jag1* phenocopies systemic *Jag1* deletion, demonstrating that a lack of Jag1 signaling from the vascular endothelium likely results in the differentiation defects, loss of vSMCs, and severe disruption of angiogenesis that can be found in *Jag1* mutants (Benedito et al. 2009; High et al. 2008). A similar loss of vSMCs has been demonstrated in embryos with homozygous hypomorphic *Notch2* (McCright et al.

2001; Wang et al. 2012). Additionally, it has been speculated that the perivascular coverage of newly formed vessels by vSMCs and pericytes is mediated by Jag1-induced expression of integrin αvβ3, which facilitates binding to a basement membrane-specific von Willebrand factor protein (Scheppke et al. 2012). In adults, Jag1 instead functions downstream of Dll4/Notch1 signaling to stimulate maturation of vSMCs after injury through $P27^{kip1}$-mediated inhibition of proliferation (Boucher et al. 2013; Pedrosa et al. 2015; reviewed in Mašek and Andersson 2017).

Jag1 also governs angiogenesis-associated sprouting; both gain- and loss-of-function investigations in endothelial cells demonstrate that Jag1 stimulates the sprouting of new tip cells during retinal angiogenesis (High et al. 2008; reviewed in Benedito and Hellström 2013; reviewed in Mašek and Andersson 2017). Notably, balanced sprouting is achieved by Dll4-induced "high" Notch signaling and inhibition of sprouting, via suppression of VEGFR signaling in tip cells, which is antagonized in stalk endothelial cells exhibiting Jag1-mediated "low" Notch signaling (Benedito et al. 2009; Pedrosa et al. 2015; reviewed in Mašek and Andersson 2017). Although these different aspects of Jag1 and Notch2 signaling have not yet been connected to Alagille or Hajdu-Cheney syndromes, they may be of relevance for the severity of these conditions and the risk for vascular accidents, including ruptured aneurysms and bleeding (Kamath et al. 2004; reviewed in Mašek and Andersson 2017).

Roles for Notch2 and Jag1 in the Embryonic Development of Inner Ear and Eye, Tissues with Important Sensory Functions

The importance of Jag1 and Notch2 for the embryonic development of organs with sensory functions is underlined by the characteristic presence of inner ear and eye defects in individuals with Alagille syndrome. Posterior embryotoxon (an irregularity of Schwalbe's line, a benign defect that is relatively common in the general population) (Emerick et al. 1999; Ozeki et al. 1997; reviewed in Mašek and Andersson 2017) represents one of the most easily detected hallmarks of Alagille syndrome. However, it must be recognized that posterior embryotoxon is difficult to study in rodents, which instead often present with eye defects, such as iris abnormalities (Xue et al. 1999; reviewed in Mašek and Andersson 2017). *Jag1* and *Notch2* are both present in the developing lens and ciliary body (CB), and Notch2 is present in the retinal pigmented epithelium (RPE) (Le et al. 2009; Saravanamuthu et al. 2012; reviewed in Mašek and Andersson 2017). During embryonic development, the Jag1-expressing inner CB interacts with the Notch2-expressing outer CB (derived from RPE) to control proliferation and bone morphogenetic protein (BMP) signaling during CB morphogenesis (Zhou et al. 2013; reviewed in Mašek and Andersson 2017). Moreover, it was demonstrated that the ectoderm-specific deletion (using Ap2a-Cre) of *Jag1* results in arrested separation of the lens vesicle from the surface ectoderm and an arrest in lens development (Le et al. 2012; reviewed in

Mašek and Andersson 2017). In the lens, deletion of *Notch2* (via Lens-Cre) also disrupts lens differentiation (Saravanamuthu et al. 2012; reviewed in Mašek and Andersson 2017) (Table 2.1). However, this phenotype is comparable to the phenotype of the heterozygous Lens-Cre-expressing mouse strain itself (Dorà et al. 2014; reviewed in Mašek and Andersson 2017).

Additionally, the Notch signaling pathway regulates via Jag1 and Notch2 also the embryonic development of the inner ear. Notably, impaired balance and deafness have been detected in four ethylnitrosourea (ENU)-induced *Jag1* mutant mouse strains, named Slalom (Tsai et al. 2001; reviewed in Mašek and Andersson 2017), Headturner (Kiernan et al. 2001; reviewed in Mašek and Andersson 2017), Ozzy (Vrijens et al. 2006; reviewed in Mašek and Andersson 2017), and Nodder (Hansson et al. 2010) (Table 2.1). It has been reported that these characteristic phenotypes are the result of the failure of Jag1-dependent Notch signaling in defining the presumptive sensory epithelium of the ear and in maintaining an adequate ratio of proliferation between populations of hair cells and supporting cells, presumably via Hes1-dependent expression of $p27^{kip}$ (Brooker et al. 2006; Kiernan et al. 2006; Murata et al. 2009; Pan et al. 2010; reviewed in Mašek and Andersson 2017). Conversely, expression of Notch1ICD (Notch1 intracellular domain) in the developing otic vesicle results in ectopic formation of sensory and supportive cells in both the vestibule and the cochlea (Pan et al. 2010; reviewed in Mašek and Andersson 2017), a process that may be responsible for the characteristic hearing deficits present in individuals with Hajdu-Cheney syndrome (Isidor et al. 2011; reviewed in Mašek and Andersson 2017). Embryonic development of sensory organs demonstrates a dose sensitivity that is comparable with other Notch-regulated processes, wherein a carefully titrated, moderate inhibition of Notch signaling activity mediated by the glycosyltransferases lunatic fringe (Lfng) and manic fringe (Mfng) represents a border between the pro-sensory primordium of the cochlear domain and Kölliker's organ. This process takes place before the cell fate decision of the first differentiating inner hair cells and their associated supporting cells, affirming the sensitivity of this organ to even very small changes in Notch signaling intensity (Basch et al. 2016; reviewed in Mašek and Andersson 2017). Notably, truncated posterior semicircular canals and missing ampullae can be found in $Jag1^{del1/+}$ and $Foxg1^{Cre+/-}$; $Jag1^{fl/+}$ heterozygous mice (Kiernan et al. 2006; reviewed in Mašek and Andersson 2017), and the severity of the vestibular phenotype in $Jag1^{del1/+}$ mice strongly correlates with the genetic background.

Conclusions

The evolutionary highly conserved Notch pathway governs many core processes including cell fate decisions during embryonic development. A huge mountain of scientific evidence convincingly demonstrates that Notch signaling represents one of the most important pathways that regulate embryogenesis in humans. Therapeutically, targeting tissues during embryogenesis may prove difficult.

However, there are some promising first scientific findings indicating that inhibition of Notch signaling through small-molecule inhibitors or antibodies may be a promising strategy to treat various disorders (Reichrath and Reichrath 2019). In contrast, therapeutic activators of Notch signaling have proven to be more difficult to develop – but may also represent promising candidates for the treatment of many diseases (Reichrath and Reichrath 2019).

References

Afelik S, Jensen J (2013) Notch signaling in the pancreas: patterning and cell fate specification. Wiley Interdiscip Rev Dev Biol 2:531–544. https://doi.org/10.1002/wdev.99

Ahnfelt-Rønne J, Jørgensen MC, Klinck R, Jensen JN, Füchtbauer E-M, Deering T, MacDonald RJ, Wright CVE, Madsen OD, Serup P (2012) Ptf1a-mediated control of Dll1 reveals an alternative to the lateral inhibition mechanism. Development 139:33–45. https://doi.org/10.1242/dev.071761

Andersson ER, Sandberg R, Lendahl U (2011) Notch signaling: simplicity in design, versatility in function. Development 138:3593–3612

Antoniou A, Raynaud P, Cordi S, Zong Y, Tronche F, Stanger BZ, Jacquemin P, Pierreux CE, Clotman F, Lemaigre FP (2009) Intrahepatic bile ducts develop according to a new mode of tubulogenesis regulated by the transcription factor SOX9. Gastroenterology 136:2325–2333. https://doi.org/10.1053/j.gastro.2009.02.051

Apelqvist A, Li H, Sommer L, Beatus P, Anderson DJ, Honjo T, Hrabe de Angelis M, Lendahl U, Edlund H (1999) Notch signalling controls pancreatic cell differentiation. Nature 400:877–881. https://doi.org/10.1038/23716

Artavanis-Tsakonas S, Rand MD, Lake RJ (1999) Notch signaling: cell fate control and signal integration in development. Science 284(5415):770–776

Aubin-Houzelstein G (2012) Notch signaling and the developing hair follicle. Adv Exp Med Biol 727:142–160. https://doi.org/10.1007/978-1-4614-0899-4_11. Review. PubMed PMID: 22399345.

Bales CB, Kamath BM, Munoz PS, Nguyen A, Piccoli DA, Spinner NB, Horn D, Shults J, Leonard MB, Grimberg A et al (2010) Pathologic lower extremity fractures in children with Alagille syndrome. J Pediatr Gastroenterol Nutr 51:66–70. https://doi.org/10.1097/MPG.0b013e3181cb9629

Basch ML, Brown RM, Jen H-I, Semerci F, Depreux F, Edlund RK, Zhang H, Norton CR, Gridley T, Cole SE et al (2016) Fine-tuning of Notch signaling sets the boundary of the organ of Corti and establishes sensory cell fates. Elife 5:841–850. https://doi.org/10.7554/eLife.19921

Bauer RC, Laney AO, Smith R, Gerfen J, Morrissette JJD, Woyciechowski S, Garbarini J, Loomes KM, Krantz ID, Urban Z et al (2010) Jagged1 (JAG1) mutations in patients with tetralogy of fallot or pulmonic stenosis. Hum Mutat 31:594–601. https://doi.org/10.1002/humu.21231

Benedito R, Hellström M (2013) Notch as a hub for signaling in angiogenesis. Exp Cell Res 319:1281–1288. https://doi.org/10.1016/j.yexcr.2013.01.010

Benedito R, Roca C, Sörensen I, Adams S, Gossler A, Fruttiger M, Adams RH (2009) The Notch ligands Dll4 and Jagged1 have opposing effects on angiogenesis. Cell 137:1124–1135. https://doi.org/10.1016/j.cell.2009.03.025

Bhatt S, Diaz R, Trainor PA (2013) Signals and switches in Mammalian neural crest cell differentiation. Cold Spring Harb Perspect Biol 5(2):a008326. https://doi.org/10.1101/cshperspect.a008326

Blanpain C, Lowry WE, Pasolli HA, Fuchs E (2006) Canonical notch signaling functions as a commitment switch in the epidermal lineage. Genes Dev 20(21):3022–3035. https://doi.org/10.1101/gad.1477606. PMCID: PMC1620020

Boucher JM, Harrington A, Rostama B, Lindner V, Liaw L (2013) A receptor-specific function for Notch2 in mediating vascular smooth muscle cell growth arrest through cyclin-dependent kinase inhibitor 1B. Circ Res 113:975–985. https://doi.org/10.1161/CIRCRESAHA.113.301272

Boyle SC, Kim M, Valerius MT, McMahon AP, Kopan R (2011) Notch pathway activation can replace the requirement for Wnt4 and Wnt9b in mesenchymal-to-epithelial transition of nephron stem cells. Development 138:4245–4254. https://doi.org/10.1242/dev.070433

Bray SJ (2016) Notch signalling in context. Nat Rev Mol Cell Biol 17:722–735. https://doi.org/10.1038/nrm.2016.94

Brooker R, Hozumi K, Lewis J (2006) Notch ligands with contrasting functions: Jagged1 and Delta1 in the mouse inner ear. Development 133:1277–1286. https://doi.org/10.1242/dev.02284

Canalis E, Adams DJ, Boskey A, Parker K, Kranz L, Zanotti S (2013) Notch signaling in osteocytes differentially regulates cancellous and cortical bone remodeling. J Biol Chem 288:25614–25625. https://doi.org/10.1074/jbc.M113.470492

Canalis E, Schilling L, Yee S-P, Lee S-K, Zanotti S (2016) Hajdu Cheney mouse mutants exhibit osteopenia, increased osteoclastogenesis, and bone resorption. J Biol Chem 291:1538–1551. https://doi.org/10.1074/jbc.M115.685453

Cheng H-T, Kim M, Valerius MT, Surendran K, Schuster-Gossler K, Gossler A, McMahon AP, Kopan R (2007) Notch2, but not Notch1, is required for proximal fate acquisition in the mammalian nephron. Development 134:801–811. https://doi.org/10.1242/dev.02773

Chiang C, Swan RZ, Grachtchouk M, Bolinger M, Litingtung Y, Robertson EK, Cooper MK, Gaffield W, Westphal H, Beachy PA, Dlugosz AA (1999) Essential role for Sonic hedgehog during hair follicle morphogenesis. Dev Biol 205:1–9

Chung E, Deacon P, Marable S, Shin J, Park J-S (2016) Notch signaling promotes nephrogenesis by downregulating Six2. Development 143:3907–3913. https://doi.org/10.1242/dev.143503

Conlon RA, Reaume AG, Rossant J (1995) Notch1 is required for the coordinate segmentation of somites. Development 121:1533–1545

Crifasi PA, Patterson MC, Michels VV (1997) Severe Hajdu-Cheney syndrome with upper airway obstruction. Am J Med Genet 70:261–266. https://doi.org/10.1002/(SICI)1096-8628(19970613)70:3<261::AID-AJMG9>3.0.CO;2-Z

D'Amato G, Luxán G, de la Pompa JL (2016a) Notch signalling in ventricular chamber development and cardiomyopathy. FEBS J 283:4223–4237. https://doi.org/10.1111/febs.13773

D'Amato G, Luxán G, del Monte-Nieto G, Martínez-Poveda B, Torroja C, Walter W, Bochter MS, Benedito R, Cole S, Martinez F et al (2016b) Sequential Notch activation regulates ventricular chamber development. Nat Cell Biol 18:7–20. https://doi.org/10.1038/ncb3280

de la Pompa JL, Wakeham A, Correia KM, Samper E, Brown S, Aguilera RJ, Nakano T, Honjo T, Mak TW, Rossant J, Conlon RA (1997) Conservation of the Notch signalling pathway in mammalian neurogenesis. Development 124(6):1139–1148

Del Monte G, Casanova JC, Guadix JA, MacGrogan D, Burch JBE, Pérez-Pomares JM, de la Pompa JL (2011) Differential notch signaling in the epicardium is required for cardiac inflow development and coronary vessel morphogenesis. Circ Res 108:824–836. https://doi.org/10.1161/CIRCRESAHA.110.229062

Devriendt K, Dooms L, Proesmans W, de Zegher F, Eggermont E, Desmet V, Devriendt K (1996) Paucity of intrahepatic bile ducts, solitary kidney and atrophic pancreas with diabetes mellitus: atypical Alagille syndrome? Eur J Pediatr 155:87–90. https://doi.org/10.1007/BF02075756

Dexter JS (1914) The analysis of a case of continuous variation in Drosophila by a study of its linkage relations. Am Nat 48:712–758. https://doi.org/10.1086/279446

Dorà NJ, Collinson JM, Hill RE, West JD (2014) Hemizygous Le-Cre transgenic mice have severe eye abnormalities on some genetic backgrounds in the absence of LoxP Sites. PLoS One 9:e109193. https://doi.org/10.1371/journal.pone.0109193

Emerick KM, Rand EB, Goldmuntz E, Krantz ID, Spinner NB, Piccoli DA (1999) Features of Alagille syndrome in 92 patients: frequency and relation to prognosis. Hepatology 29:822–829. https://doi.org/10.1002/hep.510290331

Engler A, Zhang R, Taylor V (2018) Notch and Neurogenesis. Adv Exp Med Biol 1066:223–234. https://doi.org/10.1007/978-3-319-89512-3_11. Review. PubMed PMID:30030829

Falix FA, Weeda VB, Labruyere WT, Poncy A, de Waart DR, Hakvoort TBM, Lemaigre F, Gaemers IC, Aronson DC, Lamers WH (2014) Hepatic Notch2 deficiency leads to bile duct agenesis perinatally and secondary bile duct formation after weaning. Dev Biol 396:201–213. https://doi.org/10.1016/j.ydbio.2014.10.002

Franco SJ, Muller U (2013) Shaping our minds: stem and progenitor cell diversity in the mammalian neocortex. Neuron 77(1):19–34. https://doi.org/10.1016/j.neuron.2012.12.022

Fuchs E (2007) Scratching the surface of skin development. Nature 445(7130):834–842. https://doi.org/10.1038/nature05659. PMCID: PMC2405926

Fuentealba LC, Rompani SB, Parraguez JI, Obernier K, Romero R, Cepko CL, Alvarez-Buylla A (2015) Embryonic origin of postnatal neural stem cells. Cell 161(7):1644–1655. https://doi.org/10.1016/j.cell.2015.05.041

Fujimura S, Jiang Q, Kobayashi C, Nishinakamura R (2010) Notch2 activation in the embryonic kidney depletes nephron progenitors. J Am Soc Nephrol 21:803–810. https://doi.org/10.1681/ASN.2009040353

Furutachi S, Miya H, Watanabe T, Kawai H, Yamasaki N, Harada Y, Imayoshi I, Nelson M, Nakayama KI, Hirabayashi Y, Gotoh Y (2015) Slowly dividing neural progenitors are an embryonic origin of adult neural stem cells. Nat Neurosci 18(5):657–665. https://doi.org/10.1038/nn.3989

Gaiano N, Nye JS, Fishell G (2000) Radial glial identity is promoted by Notch1 signaling in the murine forebrain. Neuron 26(2):395–404

Garg V, Muth AN, Ransom JF, Schluterman MK, Barnes R, King IN, Grossfeld PD, Srivastava D (2005) Mutations in NOTCH1 cause aortic valve disease. Nature 437:270–274. https://doi.org/10.1038/nature03940

Gazave E, Lapébie P, Richards GS, Brunet F, Ereskovsky AV, Degnan BM, Borchiellini C, Vervoort M, Renard E (2009) Origin and evolution of the Notch signalling pathway: an overview from eukaryotic genomes. BMC Evol Biol 9:249. https://doi.org/10.1186/1471-2148-9-249

Geisler F, Nagl F, Mazur PK, Lee M, Zimber-Strobl U, Strobl LJ, Radtke F, Schmid RM, Siveke JT (2008) Liver-specific inactivation of Notch2, but not Notch1, compromises intrahepatic bile duct development in mice. Hepatology 48:607–616. https://doi.org/10.1002/hep.22381

Golson ML, Le Lay J, Gao N, Brämswig N, Loomes KM, Oakey R, May CL, White P, Kaestner KH (2009a) Jagged1 is a competitive inhibitor of Notch signaling in the embryonic pancreas. Mech Dev 126:687–699. https://doi.org/10.1016/j.mod.2009.05.005

Golson ML, Loomes KM, Oakey R, Kaestner KH (2009b) Ductal malformation and pancreatitis in mice caused by conditional Jag1 deletion. Gastroenterology 136:1761–1771.e1. https://doi.org/10.1053/j.gastro.2009.01.040

Gordillo M, Evans T, Gouon-Evans V (2015) Orchestrating liver development. Development 142:2094–2108. https://doi.org/10.1242/dev.114215

Gotz M, Huttner WB (2005) The cell biology of neurogenesis. Nat Rev Mol Cell Biol 6(10):777–788. https://doi.org/10.1038/nrm1739

Greig LC, Woodworth MB, Galazo MJ, Padmanabhan H, Macklis JD (2013) Molecular logic of neocortical projection neuron specification, development and diversity. Nat Rev Neurosci 14(11):755–769. https://doi.org/10.1038/nrn3586

Gridley T (2003) Notch signaling and inherited disease syndromes. Hum Mol Genet 12:R9–R13. https://doi.org/10.1093/hmg/ddg052

Grieskamp T, Rudat C, Ludtke TH-W, Norden J, Kispert A (2011) Notch signaling regulates smooth muscle differentiation of epicardium-derived cells. Circ Res 108:813–823. https://doi.org/10.1161/CIRCRESAHA.110.228809

Gritli-Linde A, Hallberg K, Harfe BD, Reyahi A, Kannius-Janson M, Nilsson J, Cobourne MT, Sharpe PT, McMahon AP, Linde A (2007) Abnormal hair development and apparent follicular transformation to mammary gland in the absence of hedgehog signaling. Dev Cell 12:99–112

Guo C, Eckler MJ, McKenna WL, McKinsey GL, Rubenstein JL, Chen B (2013) Fezf2 expression identifies a multipotent progenitor for neocortical projection neurons, astrocytes, and oligodendrocytes. Neuron 80(5):1167–1174. https://doi.org/10.1016/j.neuron.2013.09.037

Hamada Y, Kadokawa Y, Okabe M, Ikawa M, Coleman JR, Tsujimoto Y (1999) Mutation in ankyrin repeats of the mouse Notch2 gene induces early embryonic lethality. Development 126:3415–3424

Hansson EM, Lanner F, Das D, Mutvei A, Marklund U, Ericson J, Farnebo F, Stumm G, Stenmark H, Andersson ER et al (2010) Control of Notch-ligand endocytosis by ligand-receptor interaction. J Cell Sci 123:2931–2942. https://doi.org/10.1242/jcs.073239

Hatakeyama J, Tomita K, Inoue T, Kageyama R (2001) Roles of homeobox and bHLH genes in specification of a retinal cell type. Development 128(8):1313–1322

Hatakeyama J, Bessho Y, Katoh K, Ookawara S, Fujioka M, Guillemot F, Kageyama R (2004) Hes genes regulate size, shape and histogenesis of the nervous system by control of the timing of neural stem cell differentiation. Development 131(22):5539–5550. https://doi.org/10.1242/dev.01436

High FA, Lu MM, Pear WS, Loomes KM, Kaestner KH, Epstein JA (2008) Endothelial expression of the Notch ligand Jagged1 is required for vascular smooth muscle development. Proc Natl Acad Sci U S A 105:1955–1959. https://doi.org/10.1073/pnas.0709663105

Hill CR, Yuasa M, Schoenecker J, Goudy SL (2014) Jagged1 is essential for osteoblast development during maxillary ossification. Bone 62:10–21. https://doi.org/10.1016/j.bone.2014.01.019

Hilton MJ, Tu X, Wu X, Bai S, Zhao H, Kobayashi T, Kronenberg HM, Teitelbaum SL, Ross FP, Kopan R et al (2008) Notch signaling maintains bone marrow mesenchymal progenitors by suppressing osteoblast differentiation. Nat Med 14:306–314. https://doi.org/10.1038/nm1716

Hirata H, Yoshiura S, Ohtsuka T, Bessho Y, Harada T, Yoshikawa K, Kageyama R (2002) Oscillatory expression of the bHLH factor Hes1 regulated by a negative feedback loop. Science 298(5594):840–843. https://doi.org/10.1126/science.1074560

Hofmann JJ, Zovein AC, Koh H, Radtke F, Weinmaster G, Iruela-Arispe ML (2010) Jagged1 in the portal vein mesenchyme regulates intrahepatic bile duct development: insights into Alagille syndrome. Development 137:4061–4072. https://doi.org/10.1242/dev.052118

Hofmann JJ, Briot A, Enciso J, Zovein AC, Ren S, Zhang ZW, Radtke F, Simons M, Wang Y, Iruela-Arispe ML (2012) Endothelial deletion of murine Jag1 leads to valve calcification and congenital heart defects associated with Alagille syndrome. Development 139:4449–4460. https://doi.org/10.1242/dev.084871

Honjo T (1996) The shortest path from the surface to the nucleus: RBP-J kappa/Su(H) transcription factor. Genes Cells 1(1):1–9

Humphreys R, Zheng W, Prince LS, Qu X, Brown C, Loomes K, Huppert SS, Baldwin S, Goudy S (2012) Cranial neural crest ablation of Jagged1 recapitulates the craniofacial phenotype of Alagille syndrome patients. Hum Mol Genet 21:1374–1383. https://doi.org/10.1093/hmg/ddr575

Isidor B, Lindenbaum P, Pichon O, Bézieau S, Dina C, Jacquemont S, Martin-Coignard D, Thauvin-Robinet C, Le Merrer M, Mandel J-L et al (2011) Truncating mutations in the last exon of NOTCH2 cause a rare skeletal disorder with osteoporosis. Nat Genet 43:306–308. https://doi.org/10.1038/ng.778

Jensen J, Pedersen EE, Galante P, Hald J, Heller RS, Ishibashi M, Kageyama R, Guillemot F, Serup P, Madsen OD (2000) Control of endodermal endocrine development by Hes-1. Nat Genet 24:36–44. https://doi.org/10.1038/71657

Jiang X, Rowitch DH, Soriano P, McMahon AP, Sucov HM (2000) Fate of the mammalian cardiac neural crest. Development 127:1607–1616

Joutel A, Corpechot C, Ducros A, Vahedi K, Chabriat H, Mouton P, Alamowitch S, Domenga V, Cécillion M, Maréchal E et al (1996) Notch3 mutations in CADASIL, a hereditary adult-onset condition causing stroke and dementia. Nature 383:707–710. https://doi.org/10.1038/383707a0

Kageyama R, Ohtsuka T, Kobayashi T (2007) The Hes gene family: repressors and oscillators that orchestrate embryogenesis. Development 134(7):1243–1251. https://doi.org/10.1242/dev.000786

Kamath BM, Spinner NB, Emerick KM, Chudley AE, Booth C, Piccoli DA, Krantz ID (2004) Vascular anomalies in Alagille Syndrome: a significant cause of morbidity and mortality. Circulation 109:1354–1358. https://doi.org/10.1161/01.CIR.0000121361.01862.A4

Kamath BM, Piccoli DA, Magee JC, Sokol RJ (2012) Pancreatic insufficiency is not a prevalent problem in alagille syndrome. J Pediatr Gastroenterol Nutr 55:612–614. https://doi.org/10.1097/MPG.0b013e31825eff61

Kamath BM, Spinner NB, Rosenblum ND (2013) Renal involvement and the role of Notch signalling in Alagille syndrome. Nat Rev Nephrol 9:409–418. https://doi.org/10.1038/nrneph.2013.102

Kazanis I, Lathia J, Moss L, ffrench-Constant C (2008) The neural stem cell microenvironment. In: StemBook. Harvard Stem Cell Institute, Cambridge. https://doi.org/10.3824/stembook.1.15.1

Kidd S, Kelley MR, Young MW (1986) Sequence of the notch locus of Drosophila melanogaster: relationship of the encoded protein to mammalian clotting and growth factors. Mol Cell Biol 6(9):3094–3108

Kiernan AE, Ahituv N, Fuchs H, Balling R, Avraham KB, Steel KP, Hrabé de Angelis M (2001) The Notch ligand Jagged1 is required for inner ear sensory development. Proc Natl Acad Sci U S A 98:3873–3878. https://doi.org/10.1073/pnas.071496998

Kiernan AE, Xu J, Gridley T (2006) The Notch ligand JAG1 is required for sensory progenitor development in the mammalian inner ear. PLoS Genet 2:e4. https://doi.org/10.1371/journal.pgen.0020004

Kiernan AE, Li R, Hawes NL, Churchill GA, Gridley T (2007) Genetic background modifies inner ear and eye phenotypes of Jag1 heterozygous mice. Genetics 177:307–311. https://doi.org/10.1534/genetics.107.075960

Kodama Y, Hijikata M, Kageyama R, Shimotohno K, Chiba T (2004) The role of notch signaling in the development of intrahepatic bile ducts. Gastroenterology 127:1775–1786. https://doi.org/10.1053/j.gastro.2004.09.004

Kopan R, Ilagan MXG (2009) The canonical Notch signaling pathway: unfolding the activation mechanism. Cell 137:216–233. https://doi.org/10.1016/j.cell.2009.03.045

Kopan R, Chen S, Liu Z (2014) Alagille, Notch, and robustness: why duplicating systems does not ensure redundancy. Pediatr Nephrol 29:651–657. https://doi.org/10.1007/s00467-013-2661-y

Kung AWC, Xiao S-M, Cherny S, Li GHY, Gao Y, Tso G, Lau KS, Luk KDK, Liu J-M, Cui B et al (2010) Association of JAG1 with bone mineral density and osteoporotic fractures: a genome-wide association study and follow-up replication studies. Am J Hum Genet 86:229–239. https://doi.org/10.1016/j.ajhg.2009.12.014

Le TT, Conley KW, Brown NL (2009) Jagged 1 is necessary for normal mouse lens formation. Dev Biol 328:118–126. https://doi.org/10.1016/j.ydbio.2009.01.015

Le TT, Conley KW, Mead TJ, Rowan S, Yutzey KE, Brown NL (2012) Requirements for Jag1-Rbpj mediated Notch signaling during early mouse lens development. Dev Dyn 241:493–504. https://doi.org/10.1002/dvdy.23739

Li L, Krantz ID, Deng Y, Genin A, Banta AB, Collins CC, Qi M, Trask BJ, Kuo WL, Cochran J et al (1997) Alagille syndrome is caused by mutations in human Jagged1, which encodes a ligand for Notch1. Nat Genet 16:243–251. https://doi.org/10.1038/ng0797-243

Li X-Y, Zhai W-J, Teng C-B (2016) Notch signaling in pancreatic development. Int J Mol Sci 17:48. https://doi.org/10.3390/ijms17010048

Lindström NO, Lawrence MLM, Burn SSF, Johansson JA, Bakker EERM, Ridgway RA, Chang C-H, Karolak MJM, Oxburgh L, Headon DJ et al (2015) Integrated β-catenin, BMP, PTEN, and Notch signalling patterns the nephron. Elife 3:e04000. https://doi.org/10.7554/elife.04000

Liu XS, Chopp M, Zhang RL, Tao T, Wang XL, Kassis H, Hozeska-Solgot A, Zhang L, Chen C, Zhang ZG (2011) MicroRNA profiling in subventricular zone after stroke: MiR-124a regulates proliferation of neural progenitor cells through Notch signaling pathway. PLoS One 6(8):e23461. https://doi.org/10.1371/journal.pone.0023461

Liu Z, Chen S, Boyle S, Zhu Y, Zhang A, Piwnica-Worms DR, Ilagan MXG, Kopan R (2013) The extracellular domain of notch2 increases its cell-surface abundance and ligand responsiveness during kidney development. Dev Cell 25:585–598. https://doi.org/10.1016/j.devcel.2013.05.022

Loomes KM, Russo P, Ryan M, Nelson A, Underkoffler L, Glover C, Fu H, Gridley T, Kaestner KH, Oakey RJ (2007) Bile duct proliferation in liver-specific Jag1 conditional knockout mice: effects of gene dosage. Hepatology 45:323–330. https://doi.org/10.1002/hep.21460

Lowell S, Jones P, Le Roux I, Dunne J, Watt FM (2000) Stimulation of human epidermal differentiation by delta-notch signalling at the boundaries of stem-cell clusters. Curr Biol 10:491–500

Lupo G, Harris WA, Lewis KE (2006) Mechanisms of ventral patterning in the vertebrate nervous system. Nat Rev Neurosci 7(2):103–114. https://doi.org/10.1038/nrn1843

Lutolf S, Radtke F, Aguet M, Suter U, Taylor V (2002) Notch1 is required for neuronal and glial differentiation in the cerebellum. Development 129(2):373–385

Luxán G, Casanova JC, Martínez-Poveda B, Prados B, D'Amato G, MacGrogan D, Gonzalez-Rajal A, Dobarro D, Torroja C, Martinez F et al (2013) Mutations in the NOTCH pathway regulator MIB1 cause left ventricular noncompaction cardiomyopathy. Nat Med 19:193–201. https://doi.org/10.1038/nm.3046

Luxán G, D'Amato G, MacGrogan D, de la Pompa JL (2016) Endocardial Notch signaling in cardiac development and disease. Circ Res 118:e1–e18. https://doi.org/10.1161/CIRCRESAHA.115.305350

MacGrogan D, D'Amato G, Travisano S, Martinez-Poveda B, Luxán G, Del Monte-Nieto G, Papoutsi T, Sbroggio M, Bou V, Gomez-Del Arco P et al (2016) Sequential ligand-dependent Notch signaling activation regulates valve primordium formation and morphogenesis. Circ Res 118:1480–1497. https://doi.org/10.1161/CIRCRESAHA.115.308077

Manderfield LJ, High FA, Engleka KA, Liu F, Li L, Rentschler S, Epstein JA (2012) Notch activation of Jagged1 contributes to the assembly of the arterial wall. Circulation 125:314–323. https://doi.org/10.1161/CIRCULATIONAHA.111.047159

Marin O (2013) Cellular and molecular mechanisms controlling the migration of neocortical interneurons. Eur J Neurosci 38(1):2019–2029. https://doi.org/10.1111/ejn.12225

Mason HA, Rakowiecki SM, Gridley T, Fishell G (2006) Loss of notch activity in the developing central nervous system leads to increased cell death. Dev Neurosci 28(1–2):49–57. https://doi.org/10.1159/000090752

Masamizu Y, Ohtsuka T, Takashima Y, Nagahara H, Takenaka Y, Yoshikawa K, Okamura H, Kageyama R (2006) Real-time imaging of the somite segmentation clock: revelation of unstable oscillators in the individual presomitic mesoderm cells. Proc Natl Acad Sci U S A 103(5):1313–1318. https://doi.org/10.1073/pnas.0508658103

Mašek J, Andersson ER (2017) The developmental biology of genetic Notch disorders. Development 2017(144):1743–1763. https://doi.org/10.1242/dev.148007

Massi D, Panelos J (2012) Notch signaling and the developing skin epidermis. Adv Exp Med Biol 727:131–141. https://doi.org/10.1007/978-1-4614-0899-4_10. Review. PubMed PMID: 22399344.

McCright B, Gao X, Shen L, Lozier J, Lan Y, Maguire M, Herzlinger D, Weinmaster G, Jiang R, Gridley T (2001) Defects in development of the kidney, heart and eye vasculature in mice homozygous for a hypomorphic Notch2 mutation. Development 128:491–502

McCright B, Lozier J, Gridley T (2002) A mouse model of Alagille syndrome: Notch2 as a genetic modifier of Jag1 haploinsufficiency. Development 129:1075–1082

McCright B, Lozier J, Gridley T (2006) Generation of new Notch2 mutant alleles. Genesis 44:29–33. https://doi.org/10.1002/gene.20181

Meester JAN, Verstraeten A, Alaerts M, Schepers D, Van Laer L, Loeys BL (2019) Overlapping but distinct roles for NOTCH receptors in human cardiovascular disease. Clin Genet 95(1):85–94. https://doi.org/10.1111/cge.13382. Epub 018 Jun 10. Review. PubMed PMID: 29767458.

Morgan TH (1917) The theory of the gene. Am Nat 19:309–310. https://doi.org/10.1086/279629

Morgan T (1928) The theory of the gene (revised ed. 1928). Yale University Press, New Haven, pp 77–81

Moriyama M, Durham AD, Moriyama H, Hasegawa K, Nishikawa S, Radtke F, Osawa M (2008) Multiple roles of Notch signaling in the regulation of epidermal development. Dev Cell 14:594–604

Murata J, Ohtsuka T, Tokunaga A, Nishiike S, Inohara H, Okano H, Kageyama R (2009) Notch-Hes1 pathway contributes to the cochlear prosensory formation potentially through the transcriptional down-regulation of $p27^{Kip1}$. J Neurosci Res 87:3521–3534. https://doi.org/10.1002/jnr.22169

Murtaugh LC, Stanger BZ, Kwan KM, Melton DA (2003) Notch signaling controls multiple steps of pancreatic differentiation. Proc Natl Acad Sci U S A 100:14920–14925. https://doi.org/10.1073/pnas.2436557100

Nakhai H, Siveke JT, Mendoza-Torres L, Mazur PK, Algül H, Radtke F, Strobl L, Zimber-Strobl U, Schmid RM (2008) Conditional ablation of Notch signaling in pancreatic development. Development 135:2757–2765. https://doi.org/10.1242/dev.013722

Nguyen BC, Lefort K, Mandinova A, Antonini D, Devgan V, Della Gatta G, Koster MI, Zhang Z, Wang J, Tommasi di Vignano A et al (2006) Cross-regulation between Notch and p63 in keratinocyte commitment to differentiation. Genes Dev 20:1028–1042

Nicola Z, Fabel K, Kempermann G (2015) Development of the adult neurogenic niche in the hippocampus of mice. Front Neuroanat 9:53. https://doi.org/10.3389/fnana.2015.00053

Nobta M, Tsukazaki T, Shibata Y, Xin C, Moriishi T, Sakano S, Shindo H, Yamaguchi A (2005) Critical regulation of bone morphogenetic protein-induced osteoblastic differentiation by Delta1/Jagged1-activated Notch1 signaling. J Biol Chem 280:15842–15848. https://doi.org/10.1074/jbc.M412891200

Noctor SC, Martinez-Cerdeno V, Ivic L, Kriegstein AR (2004) Cortical neurons arise in symmetric and asymmetric division zones and migrate through specific phases. Nat Neurosci 7(2):136–144. https://doi.org/10.1038/nn1172

Noctor SC, Martinez-Cerdeno V, Kriegstein AR (2007) Contribution of intermediate progenitor cells to cortical histogenesis. Arch Neurol 64(5):639–642. https://doi.org/10.1001/archneur.64.5.639

Oda T, Elkahloun AG, Pike BL, Okajima K, Krantz ID, Genin A, Piccoli DA, Meltzer PS, Spinner NB, Collins FS et al (1997) Mutations in the human Jagged1 gene are responsible for Alagille syndrome. Nat Genet 16:235–242. https://doi.org/10.1038/ng0797-235

Ohtsuka T, Ishibashi M, Gradwohl G, Nakanishi S, Guillemot F, Kageyama R (1999) Hes1 and Hes5 as notch effectors in mammalian neuronal differentiation. EMBO J 18(8):2196–2207. https://doi.org/10.1093/emboj/18.8.2196

Okuyama R, Nguyen BC, Talora C, Ogawa E, Tommasi di Vignano A, Lioumi M, Chiorino G, Tagami H, Woo M, Dotto GP (2004) High commitment of embryonic keratinocytes to terminal differentiation through a Notch1-caspase 3 regulatory mechanism. Dev Cell 6:551–562

Oro AE, Higgins K (2003) Hair cycle regulation of Hedgehog signal reception. Dev Biol 255:238–248

Ozeki H, Shirai S, Majima A, Sano M, Ikeda K (1997) Clinical evaluation of posterior embryotoxon in one institution. Jpn J Ophthalmol 41:422–425. https://doi.org/10.1016/S0021-5155(97)00080-4

Pan W, Jin Y, Stanger B, Kiernan AE (2010) Notch signaling is required for the generation of hair cells and supporting cells in the mammalian inner ear. Proc Natl Acad Sci U S A 107:15798–15803. https://doi.org/10.1073/pnas.1003089107

Pedrosa A-R, Trindade A, Fernandes A-C, Carvalho C, Gigante J, Tavares AT, Diéguez-Hurtado R, Yagita H, Adams RH, Duarte A (2015) Endothelial jagged1 antagonizes Dll4 regulation of endothelial branching and promotes vascular maturation downstream of Dll4/Notch1. Arterioscler Thromb Vasc Biol 35:1134–1146. https://doi.org/10.1161/ATVBAHA.114.304741

Perez-Pomares JM, de la Pompa JL (2011) Signaling during epicardium and coronary vessel development. Circ Res 109:1429–1442. https://doi.org/10.1161/CIRCRESAHA.111.245589

Reichrath J, Reichrath S (2019) Notch signalling and inherited diseases: challenge and promise. Adv Exp Med Biol, in press.

Richards GS, Degnan BM (2009) The dawn of developmental signaling in the metazoa. Cold Spring Harb Symp Quant Biol 74:81–90. https://doi.org/10.1101/sqb.2009.74.028

Riely CA, Cotlier E, Jensen PS, Klatskin G (1979) Arteriohepatic dysplasia: a benign syndrome of intrahepatic cholestasis with multiple organ involvement. Ann Intern Med 91:520–527. https://doi.org/10.7326/0003-4819-91-4-520

Rolando C, Taylor V (2014) Neural stem cell of the hippocampus: development, physiology regulation, and dysfunction in disease. Curr Top Dev Biol 107:183–206. https://doi.org/10.1016/B978-0-12-416022-4.00007-X

Rovner AJ, Schall JI, Jawad AF, Piccoli DA, Stallings VA, Mulberg AE, Zemel BS (2002) Rethinking growth failure in Alagille syndrome: the role of dietary intake and steatorrhea. J Pediatr Gastroenterol Nutr 35:495–502. https://doi.org/10.1097/00005176-200210000-00007

Saravanamuthu SS, Le TT, Gao CY, Cojocaru RI, Pandiyan P, Liu C, Zhang J, Zelenka PS, Brown NL (2012) Conditional ablation of the Notch2 receptor in the ocular lens. Dev Biol 362:219–229. https://doi.org/10.1016/j.ydbio.2011.11.011

Sauka-Spengler T, Bronner M (2010) Snapshot: neural crest. Cell 143(3):486–486. https://doi.org/10.1016/j.cell.2010.10.025

Scheppke L, Murphy EA, Zarpellon A, Hofmann JJ, Merkulova A, Shields DJ, Weis SM, Byzova TV, Ruggeri ZM, Iruela-arispe ML et al (2012) Notch promotes vascular maturation by inducing integrin-mediated smooth muscle cell adhesion to the endothelial basement membrane. Blood 119:2149–2158. https://doi.org/10.1182/blood-2011-04-348706

Sedmera D (2011) Function and form in the developing cardiovascular system. Cardiovasc Res 91:252–259. https://doi.org/10.1093/cvr/cvr062

Shih HP, Kopp JL, Sandhu M, Dubois CL, Seymour PA, Grapin-Botton A, Sander M (2012) A Notch-dependent molecular circuitry initiates pancreatic endocrine and ductal cell differentiation. Development 139:2488–2499. https://doi.org/10.1242/dev.078634

Shimojo H, Ohtsuka T, Kageyama R (2008) Oscillations in notch signaling regulate maintenance of neural progenitors. Neuron 58(1):52–64. https://doi.org/10.1016/j.neuron.2008.02.014

Shimojo H, Ohtsuka T, Kageyama R (2011) Dynamic expression of notch signaling genes in neural stem/progenitor cells. Front Neurosci 5:78. https://doi.org/10.3389/fnins.2011.00078

Stump G, Durrer A, Klein AL, Lutolf S, Suter U, Taylor V (2002) Notch1 and its ligands Delta-like and Jagged are expressed and active in distinct cell populations in the postnatal mouse brain. Mech Dev 114(1–2):153–159

Tam PP, Loebel DA (2007) Gene function in mouse embryogenesis: get set for gastrulation. Nat Rev Genet 8(5):368–381. https://doi.org/10.1038/nrg2084

Telley L, Govindan S, Prados J, Stevant I, Nef S, Dermitzakis E, Dayer A, Jabaudon D (2016) Sequential transcriptional waves direct the differentiation of newborn neurons in the mouse neocortex. Science 351(6280):1443–1446. https://doi.org/10.1126/science.aad8361

Thakurdas SM, Lopez MF, Kakuda S, Fernandez-Valdivia R, Zarrin-Khameh N, Haltiwanger RS, Jafar-Nejad H (2016) Jagged1 heterozygosity in mice results in a congenital cholangiopathy which is reversed by concomitant deletion of one copy of Poglut1 (Rumi). Hepatology 63:550–565. https://doi.org/10.1002/hep.28024

Theodoris CV, Li M, White MP, Liu L, He D, Pollard KS, Bruneau BG, Srivastava D (2015) Human disease modeling reveals integrated transcriptional and epigenetic mechanisms of NOTCH1 haploinsufficiency. Cell 160:1072–1086. https://doi.org/10.1016/j.cell.2015.02.035

Tsai H, Hardisty RE, Rhodes C, Kiernan AE, Roby P, Tymowska-Lalanne Z, Mburu P, Rastan S, Hunter AJ, Brown SDM et al (2001) The mouse slalom mutant demonstrates a role for Jagged1 in neuroepithelial patterning in the organ of Corti. Hum Mol Genet 10:507–512. https://doi.org/10.1093/hmg/10.5.507

Varadkar P, Kraman M, Despres D, Ma G, Lozier J, McCright B (2008) Notch2 is required for the proliferation of cardiac neural crest-derived smooth muscle cells. Dev Dyn 237:1144–1152. https://doi.org/10.1002/dvdy.21502

Vrijens K, Thys S, De Jeu MT, Postnov AA, Pfister M, Cox L, Zwijsen A, Van Hoof V, Mueller M, De Clerck NM et al (2006) Ozzy, a Jag1 vestibular mouse mutant, displays characteristics of Alagille syndrome. Neurobiol Dis 24:28–40. https://doi.org/10.1016/j.nbd.2006.04.016

Walter TJ, Vanderpool C, Cast AE, Huppert SS (2014) Intrahepatic bile duct regeneration in mice does not require Hnf6 or Notch signaling through Rbpj. Am J Pathol 184:1479–1488. https://doi.org/10.1016/j.ajpath.2014.01.030

Wang X, Pasolli HA, Williams T, Fuchs E (2008) AP-2 factors act in concert with Notch to orchestrate terminal differentiation in skin epidermis. J Cell Biol 183:37–48

Wang Q, Zhao N, Kennard S, Lilly B (2012) Notch2 and notch3 function together to regulate vascular smooth muscle development. PLoS One 7:e37365. https://doi.org/10.1371/journal.pone.0037365

Weller M, Krautler N, Mantei N, Suter U, Taylor V, (2006) Jagged1 ablation results in cerebellar granule, cell migration defects and depletion of Bergmann glia. Dev Neurosci 28(1–2):70–80. https://doi.org/10.1159/000090754

Wharton KA, Johansen KM, Xu T, Artavanis-Tsakonas S (1985) Nucleotide sequence from the neurogenic locus notch implies a gene product that shares homology with proteins containing EGF-like repeats. Cell 43(3 Pt 2):567–581

Xue Y, Gao X, Lindsell CE, Norton CR, Chang B, Hicks C, Gendron-Maguire M, Rand EB, Weinmaster G, Gridley T (1999) Embryonic lethality and vascular defects in mice lacking the Notch ligand Jagged1. Hum Mol Genet 8:723–730. https://doi.org/10.1093/hmg/8.5.723

Yang J, Bücker S, Jungblut B, Böttger T, Cinnamon Y, Tchorz J, Müller M, Bettler B, Harvey R, Sun Q-Y et al (2012) Inhibition of Notch2 by numb/Numblike controls myocardial compaction in the heart. Cardiovasc Res 96:276–285. https://doi.org/10.1093/cvr/cvs250

Youngstrom DW, Dishowitz MI, Bales CB, Carr E, Mutyaba PL, Kozloff KM, Shitaye H, Hankenson KD, Loomes KM (2016) Jagged1 expression by osteoblast-lineage cells regulates trabecular bone mass and periosteal expansion in mice. Bone 91:64–74. https://doi.org/10.1016/j.bone.2016.07.006

Zanotti S, Canalis E (2013) Notch signaling in skeletal health and disease. Eur J Endocrinol 168(6):R95–R103. https://doi.org/10.1530/EJE-13-0115. Print 2013 Jun. Review. PubMed PMID: 23554451; PubMed Central PMCID: PMC4501254.

Zanotti S, Canalis E (2016) Notch signaling and the skeleton. Endocr Rev 37:223–253. https://doi.org/10.1210/er.2016-1002

Zhou Y, Tanzie C, Yan Z, Chen S, Duncan M, Gaudenz K, Li H, Seidel C, Lewis B, Moran A et al (2013) Notch2 regulates BMP signaling and epithelial morphogenesis in the ciliary body of the mouse eye. Proc Natl Acad Sci U S A 110:8966–8971. https://doi.org/10.1073/pnas.1218145110

Zong Y, Panikkar A, Xu J, Antoniou A, Raynaud P, Lemaigre F, Stanger BZ (2009) Notch signaling controls liver development by regulating biliary differentiation. Development 136:1727–1739. https://doi.org/10.1242/dev.029140

Chapter 3
The Five Faces of Notch Signalling During *Drosophila melanogaster* Embryonic CNS Development

Shahrzad Bahrampour and Stefan Thor

Abstract During central nervous system (CNS) development, a complex series of events play out, starting with the establishment of neural progenitor cells, followed by their asymmetric division and formation of lineages and the differentiation of neurons and glia. Studies in the *Drosophila melanogaster* embryonic CNS have revealed that the Notch signal transduction pathway plays at least five different and distinct roles during these events. Herein, we review these many faces of Notch signalling and discuss the mechanisms that ensure context-dependent and compartment-dependent signalling. We conclude by discussing some outstanding issues regarding Notch signalling in this system, which likely have bearing on Notch signalling in many species.

Keywords CNS development · Notch signalling

Introduction

The Notch signalling pathway is well conserved throughout evolution and controls a number of biological events. Research on Notch pathway dates back to more than a century, when the Notch mutant was first identified in *Drosophila melanogaster* [described in Poulson (1937)]. The importance of the Notch pathway is increasing, regarding its role both in development and in disease biology. Notch signalling is

S. Bahrampour (✉)
The Hospital for Sick Children, Peter Gilgan Center for Research and Learning, Toronto, ON, Canada

Karolinska Institutet, Department of Cell and Molecular Biology (CMB), Stockholm, Sweden
e-mail: shahrzad.bahrampour@sickkids.ca

S. Thor (✉)
School of Biomedical Sciences, University of Queensland, St Lucia, QLD, Australia
e-mail: s.thor@uq.edu.au

J. Reichrath, S. Reichrath (eds.), *Notch Signaling in Embryology and Cancer*, Advances in Experimental Medicine and Biology 1218,
https://doi.org/10.1007/978-3-030-34436-8_3

essential for many different organs and plays an instrumental role during the development of the central nervous system (CNS). The powerful molecular and genetic tools developed in *Drosophila* and the relative simplicity of its CNS, when compared to mammals, have made it an invaluable model system for developmental neurobiology and for decoding Notch signalling (Allan and Thor 2015; Skeath and Thor 2003). In this chapter, we will focus on the contributions of this model system to our understanding of the different roles of Notch signal transduction specifically during *Drosophila* embryonic CNS development.

We start by providing an insight into the molecular aspects of the pathway. Similar to many other signal transduction pathways, the Notch pathway proceeds both along a more frequently used, so-called canonical pathway, and along one (perhaps several) unusual, non-canonical routes.

We then go through five well-characterized biological roles for the Notch pathway in the *Drosophila* embryonic CNS:

First, the selection of neural progenitor cells (neuroblasts; NBs) in the neuroectoderm.

Second, the control of proliferation of daughter cells generated by NBs, the so-called Type I->0 daughter cell proliferation switch.

Third, the role of Notch during daughter cell asymmetric cell division, acting to promote different neuronal cell fates in the sibling cells.

Fourth, the role of Notch signalling in the development of the glia/axon scaffold, which involves glia to interneuron axon interactions.

Fifth, the role of Notch in motor axon pathfinding.

Intriguingly, these studies collectively reveal that the same NB, and its daughter cells, its lineage, may be involved in up to four of these five Notch-mediated signalling events during the course of embryonic CNS development. A number of asymmetrically distributed proteins and several feedback loops act to gate Notch signalling during these temporally tightly interwoven events. However, the dynamics of these events indicate that some control mechanisms likely remain undiscovered.

Against the backdrop of the pervasive role of Notch signalling during mammalian development and adult homeostasis, as well as its frequent involvement in human disease, studies in the *Drosophila* embryonic CNS have provided seminal insight into this pathway and many of these findings have had direct bearing on human biology.

Notch Canonical Molecular Pathway

The Notch mutation was first identified in *Drosophila* more than a decade ago (Poulson 1937). However, its features at the molecular level remained elusive until the 1980s (Artavanis-Tsakonas et al. 1983; Wharton et al. 1985). The Notch protein is a Type I transmembrane glycoprotein, and the Notch extracellular domain contains several epidermal growth factor (EGF)-like and cell LINeage defective-12

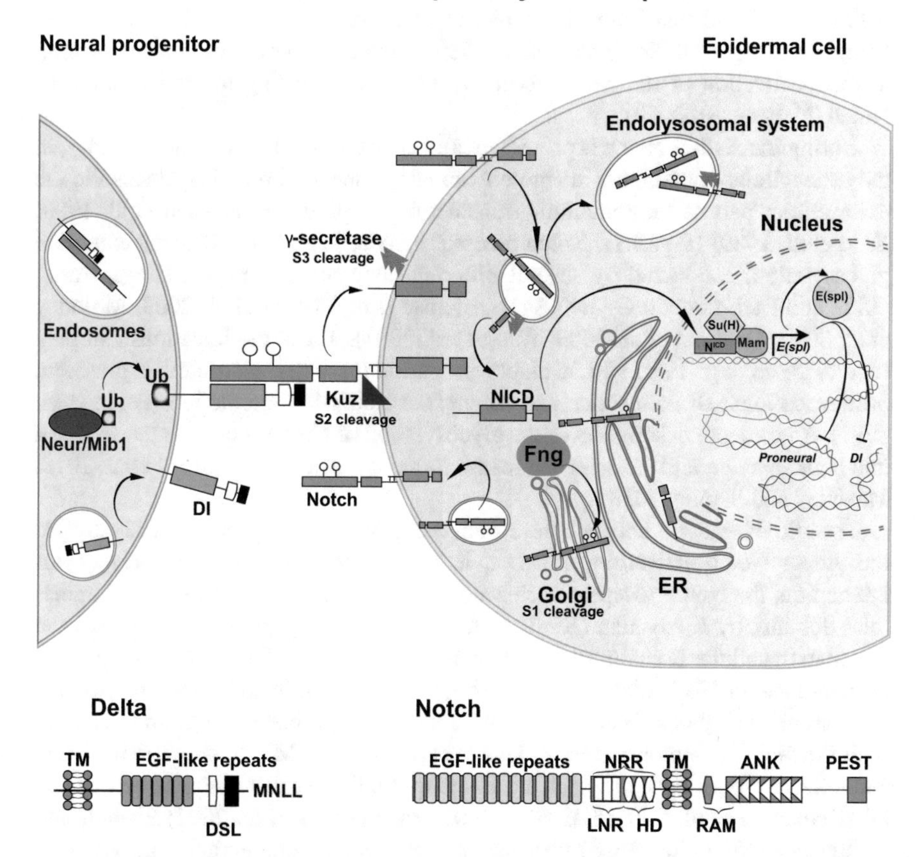

Fig. 3.1 Molecular aspects of canonical Notch signalling during *Drosophila* development. (Top) cartoon illustrating the *Drosophila* canonical Notch signalling pathway between two adjacent cells. The cell to the left is signal-sending (epidermal cell) and the one to the right is the signal-receiving cell, which is forced to an NB fate. (Bottom) The domain organization of the *Drosophila* Notch receptor and Delta ligand is illustrated with higher magnification. (See main text for details)

(LIN-12/Notch repeat) motifs (Wharton et al. 1985) (Fig. 3.1). The Notch intracellular part has cdc10/ankyrin (ANK) motifs (Zweifel et al. 2003). In *Drosophila*, two Notch ligands have been identified: Delta (Dl) and Serrate (Ser), both of which are also Type I transmembrane domain proteins containing multiple EGF repeats in the extracellular domain (Fleming et al. 1990; Thomas et al. 1991; Vassin et al. 1987).

Activation of the Notch pathway in a canonical manner occurs between two cells, i.e. a signal-receiving and a signal-sending cell (juxtacrine signalling). Generally, Notch receptor expression is broad, when compared to the expression pattern of the two ligands Dl and Ser (Bachmann and Knust 1998; Bender et al. 1993; Kopczynski and Muskavitch 1989; Thomas et al. 1991). Hence, the spatio-temporal expression pattern of Dl and Ser helps determine the place and time of

Notch activation. However, in many cases, such as in the process of lateral inhibition, it has been found that both of the two adjacent cells express both receptor and ligand. Strikingly, Dl-Notch can interact both in the canonical trans manner, leading to trans-activation of the Notch receptor, and in cis, leading to cis-inhibition of Notch (Miller et al. 2009).

During translation Notch is proteolytically cleaved (S1 cleavage) in the Golgi, in the extracellular domain, but the protein domain cleaved off remains attached to the extracellular part of the remaining transmembrane protein (Johansen et al. 1989; Kidd et al. 1989) (Fig. 3.1). Notch receptor activity is also modified by extensive glycosylation, conducted by several different enzymes, i.e. Fringe (Fng), Rumi, Ofut1 and EGF-domain O-GlcNAc transferase (Eogt) (Acar et al. 2008; Moloney et al. 2000; Okajima et al. 2003; Wang et al. 2001). These modifications can play several roles, e.g. Fng, which displays spatiotemporally restricted expression, promotes Notch-Dl connection and suppresses Notch-Ser interaction (Panin et al. 1997). An in-depth description of the glycosylation of Notch is outside the scope of this book chapter, and we refer the reader to recent reviews (Handford et al. 2018; Varshney and Stanley 2018).

Specific interaction between the Notch receptor and its ligands is necessary for the proper Notch activation; however, it is not sufficient. Upon ligand-receptor interaction, the Notch receptor is cleaved for the second time (S2) in the extracellular domain, by Kuzbanian (Kuz), an ADAM metallopeptidase, which leaves the Notch extracellular domain truncated but attached as a membrane-attached portion (Pan and Rubin 1997; Lieber et al. 2002; Qi et al. 1999). In addition, the cytoplasmic part of the ligands Dl and Ser must be mono-ubiquitinated for them to interact with the Notch receptor correctly. This is carried out by Mindbomb 1 (Mib1) and Neuralized (Neur), both of which are E3 ubiquitin ligases that ubiquitinate Ser and Dl (Deblandre et al. 2001; Lai et al. 2001; Pavlopoulos et al. 2001; Pitsouli and Delidakis 2005). Following S2 cleavage on Notch, an intramembrane protease complex, named gamma-secretase, implements the third cleavage (S3) on the Notch extracellular truncation (NEXT). Consequently, the intracellular domain of Notch (NICD or Notch-intra) is released into the cytosol [reviewed in Bray (2006) and Fortini (2009)] (Fig. 3.1).

Characteristically, NICD moves to the nucleus and partners with Suppressor of Hairless (Su(H)), a transcription factor (TF) (Bailey and Posakony 1995; Fortini and Artavanis-Tsakonas 1994; Schweisguth and Posakony 1994) and its cofactor, Mastermind (Mam) (Nam et al. 2006; Petcherski and Kimble 2000; Wilson and Kovall 2006; Wu et al. 2000), to form a transcriptional activation complex. The interaction of NICD with Su(H) in the nucleus alters the role of Su(H), from a repressor TF to an activator one. The Su(H)-Mam-NICD complex activates the expression of the enhancer of split complex (E(spl)-C) (Bailey and Posakony 1995; Jennings et al. 1994; Lecourtois and Schweisguth 1995). E(spl)-C encodes seven bHLH repressor TFs (E(spl)-m3, m5, m7, m8, mβ, mγ, mδ) (Delidakis et al. 1991; Klambt et al. 1989; Knust et al. 1992). E(spl)-C represses the expression of the proneural bHLH TFs of the achaete-scute complex (AS-C) during neurodevelopment (Nakao and Campos-Ortega 1996; Oellers et al. 1994).

The function of these proneural TFs, as well as other Notch target genes, is crucial during neurodevelopment (as explained in the sections below).

Finally, the termination of Notch signalling is not as well understood as its activation. However, one important mechanism revolves around the degradation of several components, including Notch itself and the Dl ligand (Lai 2002). Studies on mammalian systems suggest that NICD degradation necessitates NICD phosphorylation mediated by CDK8 (cyclin-dependent kinase 8), which facilitates NICD ubiquitination by the ubiquitin E3 ligase FBXW7 [reviewed in Yeh et al. (2018)]. The *Drosophila* FBXW7 orthologue Archipelago (Ago) displays Notch-related phenotypes and regulates Notch target genes (Bivik Stadler et al. 2019; Nicholson et al. 2011). However, it is unclear if *Drosophila* Ago plays the same role as the mammalian FBXW7 protein.

Notch Non-canonical Molecular Pathway

In the majority of contexts, Notch appears to function along the canonical cascade, i.e. controlling the transcription of Notch downstream target genes. However, there are exceptions. One of the most studied and the best examples of the non-canonical function of Notch is the Notch/Abl signalling pathway. Notch/Abl pathway functions in the growth and pathfinding of pioneer axons in the *Drosophila* embryonic CNS (Giniger 2012; Giniger et al. 1993). In this concept, Notch directly interacts with Disabled (Dab) and Trio proteins, the two upstream factors of the Abl tyrosine kinase pathway, to locally suppress Abl signalling (Crowner et al. 2003; Giniger 1998; Kuzina et al. 2011; Le Gall et al. 2008) (Fig. 3.2). Dab is an adaptor

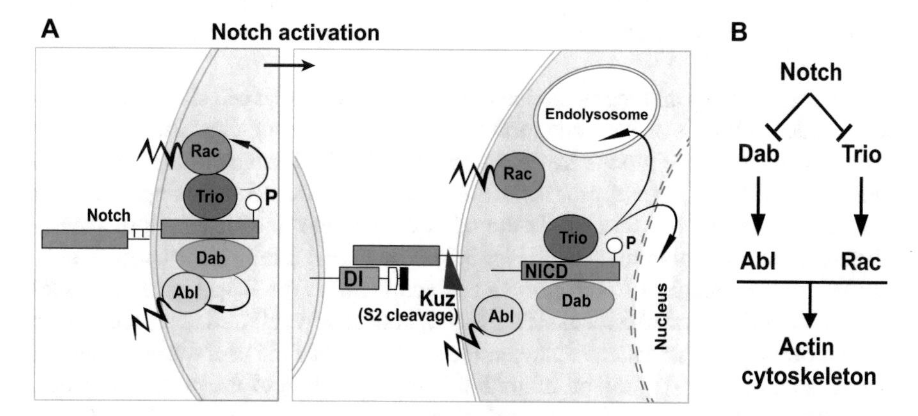

Fig. 3.2 Molecular aspects of non-canonical Notch signalling during *Drosophila* CNS development. Cartoon illustrating the non-canonical Notch signalling activation between two adjacent cells. (**a**) Notch/Abl interaction before (left) and after (right) activation of Notch. (**b**) Model depicting downstream targets of Notch in the non-canonical pathway. (See main text for details)

protein that activates Abl kinase activates by localizing Abl protein (Kannan et al. 2017; Song and Giniger 2011), and Trio is a guanine exchange factor (GEF), which acts via Rac GTPase to trigger Abl pathway (Newsome et al. 2000).

However, Notch/Abl nontraditional Notch activity requires Notch receptor-ligand interaction, including the cascades of proteolytic cleavages of Notch (S1-S3) (Fig. 3.1). Additionally, it requires Notch direct interaction with Dab and Trio, earlier, and following Notch-ligand activation. Further, Notch is required to be tyrosine-phosphorylated for its association with Dab and Trio, while canonical Notch activity does not necessitate a tyrosine-phosphorylated form of Notch (Kannan et al. 2018).

Neuroblast Selection and Delamination

The *Drosophila* CNS can be subdivided into the brain and the ventral nerve cord (VNC). The VNC originates from the two ventrolateral neurogenic regions, during early embryogenesis. The ventral neurogenic regions fuse during gastrulation and form a uniform layer of cells, named neuroectoderm. The neuroectoderm is a monolayer cell sheet that is segmented both along the anterior-posterior (A-P) and the dorsoventral (D-V) axis. Segmentation of neuroectoderm and CNS is due to the sequential action of a distinct set of genes, including segment polarity and columnar genes, which regulate (A-P) and (D-V) patterning, respectively [reviewed in Lawrence et al. (1996)]. The brain forms from the two anterior neurogenic regions and undergoes an even more complex set of patterning events (Cohen and Jurgens 1990; Hirth and Reichert 1999; Thor 1995). The brain is segmented into three regions (B1–B3), while the VNC contains several neuromeric segments, replicated along the A-P axis (S1-S3, T1-T3, A1-A10) (Urbach et al. 2003, 2016; Birkholz et al. 2013). Each segment displays a bilateral symmetry and can be divided into two equivalent hemisegments, separated by specialized midline progenitors (Wheeler et al. 2009).

After generation and segmentation of the neuroectodermal cell sheet, stereotypically spaced clusters of 5–6 cells form the so-called proneural clusters, also known as neural equivalence groups [reviewed in Bhat (1999), Skeath (1999) and Skeath and Thor (2003)]. After a process denoted lateral inhibition, one single cell per cluster is assigned NB fate, while the other cells within the cluster subsequently differentiate into epidermal cells (Fig. 3.3). The lateral inhibition process is controlled by the canonical Notch pathway, which includes the E(spl) genes (see section "Notch Canonical Molecular Pathway"), collectively referred to as neurogenic genes, because the mutants display an increase in NBs. One of the key outputs of Notch signalling during lateral inhibition is the regulation of the proneural genes, which promote neuronal fate. Proneural genes are a set of related bHLH TFs, encoded by the *acheate* (*ac*), *scute* (*sc*) and *lethal of scute* (*l(1)sc*) genes, which are located in a genomic region denoted the *achaete-scute complex* (*AS-C*) (Bhat 1999; Skeath 1999; Skeath and Thor 2003).

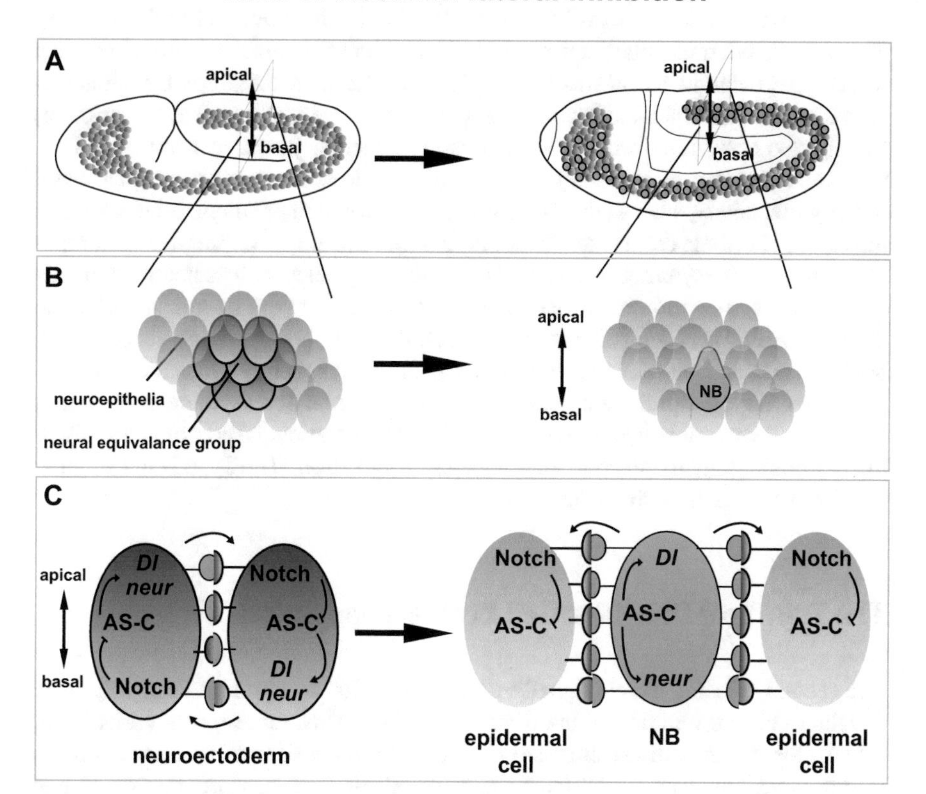

Fig. 3.3 NB selection and Notch signalling in the *Drosophila* neuroectoderm. (**a**) Lateral view of early embryos depicting NB generation within the neuroectoderm. (**b**) Magnified view of a proneural cluster within the neuroectoderm (left) and NB selection and delamination (right). (**c**) Two equipotent cells within a proneural cluster (left), which undergo lateral inhibition to select one cell, with low/no Notch activity, that acquires NB fate (right)

During NB selection, binding of Dl to the Notch receptor, and ubiquitination of Dl by the E3 ligase Neuralized (Neur), triggers cleavage of the Notch intracellular domain (NICD), followed by its transfer to the nucleus, where NICD partners with Suppressor of Hairless [Su(H)], a DNA-binding protein, and its co-factor Mastermind (Mam). This tripartite protein complex promotes the expression of the E(spl) bHLH TFs. The E(spl) genes in turn repress the expression of the *AS-C*, *Dl* and *neur* genes. Hence, activation of NICD leads to downregulation of proneural genes, thus preventing NB fate in favour of epidermal cell differentiation. In contrast, expression of proneural genes continues in the cells with low/no Notch activity within each proneural cluster, and the proneural genes will continue driving expression of *Dl* and *neur*. Therefore, cells with low Notch activity will continue presenting Dl to neighbouring cells, activating Notch in those cells and inhibiting them from acquiring NB fate.

The selected NBs will enlarge and segregate from the neuroectoderm, to move inside the embryo, in a process known as NB delamination (Doe and Technau 1993). The neuroectoderm displays apico-basal polarity, and NBs break out from this sheet of cells and delaminate towards the basal side of the neuroectoderm. Delaminating NBs abridge their apical side from the neuroectodermal cells (apical constriction) by repeated myosin pulsation and eliminating adherens junctions from their adjacent cells (Simoes et al. 2017). NBs downregulate some aspects of apico-basal polarity machinery, such as the Crumbs complex, while maintaining others, such as the Par and Scribble complexes. In addition, NBs activate the expression of components of the NB asymmetric cell division machinery, such as inscuteable, miranda and prospero. The process of delamination of NBs from the neuroectoderm, and the establishment of the NB asymmetric cell division, is outside the scope of this book chapter, and we refer to recent reviews. The connection between the canonical Notch pathway, the proneural genes and the apico-basal and asymmetric cell division machinery is not clear. Immediately after delamination, NBs commence dividing asymmetrically to generate their lineages (see section "The Type I->0 Daughter Cell Proliferation Switch") (Fig. 3.3).

The Type I->0 Daughter Cell Proliferation Switch

The second role for Notch signalling pertains to its role in controlling alternate daughter cell proliferation in the developing VNC. This, more recently identified, role for the Notch pathway emerged from a forward genetic screen, aimed at identifying genes involved in the later stages of NB5-6T lineage progression of one specific NB lineage in the thoracic VNC (Ulvklo et al. 2012).

As outlined above, in each hemisegment of the embryonic VNC, there are some 30 lateral NBs. Most, if not all, of these NBs begin neurogenesis by proliferating in the Type I daughter proliferation mode. This refers to that they generate a daughter cell, a ganglion mother cell (GMC), which divides once to generate two neurons/glia (Baumgardt et al. 2014; Boone and Doe 2008; Doe and Technau 1993). During subsequent developmental stages, many NBs switch to the Type 0 daughter proliferation mode, referring to that they generate daughter cells that differentiate directly (Baumgardt et al. 2014) (Fig. 3.4a). Two model lineages, NB5-6T, which was used for the aforementioned genetic screen, and NB3-3A, have been particularly helpful in decoding the Type I->0 daughter cell proliferation switch (Baumgardt et al. 2009, 2014; Bivik et al. 2016; Ulvklo et al. 2012). Intriguingly, detailed lineage analysis has revealed that the lineage topology is different for NB5-6T and NB3-3A: NB5-6T undergoes nine rounds of Type I proliferation, followed by five rounds of Type 0, while, in contrast, NB3-3A undergoes one round of Type I, followed by 11 rounds of Type 0 proliferation (Baumgardt et al. 2009, 2014; Bivik et al. 2016; Ulvklo et al. 2012). This means that both the Type I->0 switch and the NB cell cycle exit are precisely controlled and that these decisions are executed at different lineage progression stages in different lineages.

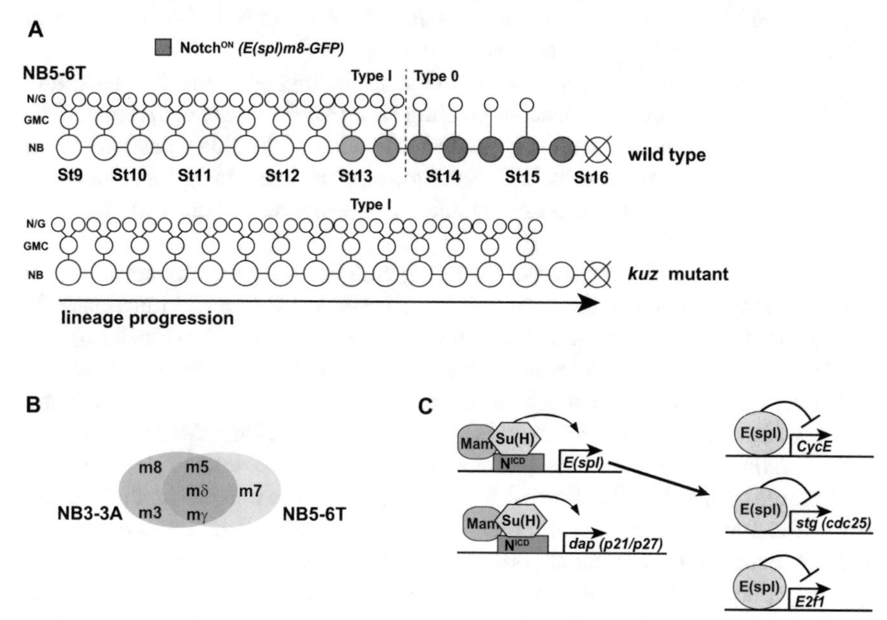

Fig. 3.4 Notch controls the Type I->0 daughter cell proliferation switch. Cellular and molecular aspects of Notch function regarding the Type I->0 switch during lineage progression. (**a**) Lineage progression of NB5-6 in the thoracic segments of the *Drosophila* VNC, during embryogenesis. Although the NB is Notch-OFF during its formation, Notch signalling, evident by *E(spl)-m8* expression, is activated in the NB during lineage progression and triggers the Type I->0 switch. In the absence of Notch signalling, e.g. in *kuz* mutants, NB5-6T fails to undergo the Type I->0 switch. (**b**) Studies have revealed differential function of seven *E(spl)* genes in the abdominal NB3-3 and thoracic NB5-6 lineages. (**c**) The anti-proliferative effect of Notch signalling during the Type I->0 switch plays out at two levels, with the NICD-Su(H)-Mam complex first activating *E(spl)* and *dap* (p21/p21) and E(spl) repressing *CycE*, *E2f1* and *stg* (cdc25)

The forward genetic screen, using NB5-6T as readout, was based upon a transgenic reporter, where the enhancer for the neuropeptide gene *FMRFamide* (*FMRFa*) was used to drive green fluorescent protein (GFP) reporter. *FMRFa* is expressed in the last-born cell in the NB5-6T lineage, in the Type 0 window, and hence mutants with additional *FMRFa-GFP* cells may reflect problems with executing the Type I->0 switch. The screen identified a number of such mutants, and two of them mapped to the Notch pathway: *neuralized* (*neur*) and *kuzbanian* (*kuz*). Analysis of these mutants was complex, because strong Notch pathway mutants also increased *FMRFa-GFP* due to the lateral inhibition effect and hence generation of additional NB5-6T lineages. This was indeed evident in the *neur* mutants. However, due to the maternal load of *kuz*, this Notch pathway mutant could be used to selectively analyse the Type I->0 switch without the confounding issue of additional lineages being generated. Moreover, using Gal4 drivers that drive expression of various Notch components after NB delamination further allowed for dissecting late from early roles of Notch signalling. These studies revealed that the canonical Notch pathway was involved in the Type I->0 switch, in both NB5-6T and NB3-3A (Ulvklo et al. 2012), as well as globally (Bivik et al. 2016) (Fig. 3.4a).

The role of Notch in the Type I->0 switch is surprising, given that NB generation is critically dependent upon Notch signalling being OFF in early ectodermal cells. However, studies reveal that there is gradual activation of the Notch pathway late in NBs, thereby triggering the switch. Hence, similar to late temporal genes (Baumgardt et al. 2014), Notch can be seen as also acting in a temporal manner with respect to the Type I->0 switch.

While the seven TFs encoded in the E(spl) complex are generally considered to act redundantly, studies of the Type I->0 switch revealed differential function of the E(spl) complex genes, with different genes acting in different NBs (Fig. 3.4b).

The canonical Notch pathway appears to, at least in part, gate the Type I->0 switch by NICD-Su(H)-Mam activating the cell cycle repressor Dacapo (mammalian *cdkn1a*, encoding $p21^{Cip1}$) expression and E(spl) repressing the cell cycle activators cyclin E, E2f1 and String (mammalian cdc25) (Fig. 3.4c).

The role of Notch in the Type I->0 switch appears to be independent of the control of the Type I daughter proliferation itself, i.e. ensuring that GMCs can divide, but only once. In most, if not all, NBs, the repetitive rounds of Type I proliferation, typically playing in the first part of NB lineages, are gated by the Prospero (Pros) factor. Inside NBs, Pros is tethered cortically and is asymmetrically distributed to the GMC at cell division. Inside the GMC, Pros enters the nucleus, where it represses *E2f1* and *CycE* gene expression, thereby triggering cell cycle exit after one division has been completed by the GMC (Hirata et al. 1995; Knoblich et al. 1995; Spana and Doe 1995; Choksi et al. 2006; Li and Vaessin 2000). The switch from Type I to Type 0 daughter cell proliferation is triggered by the activation of Dap expression in NBs. Dap acts in the daughter cell, against the backdrop of declining E2f1 and CycE levels, to block the cell entry into S-phase, and thereby stops Type 0 daughter cells from dividing (Baumgardt et al. 2014). While *pros* mutants display extensive overgrowth of daughter cells in the Type I window, there was no apparent effect observed in the Type 0 daughter cells (Ulvklo et al. 2012). Conversely, Notch pathway interference only affected Type 0 daughter cells. This prompted the investigators to generate *pros*, *kuz* double mutants, which intriguingly displayed overgrowth of both Type I and Type 0 daughters and hence massive lineage expansion.

In addition to the canonical Notch pathway, the aforementioned genetic screen (Ulvklo et al. 2012) also identified other genes that may relate to Notch signalling. These included *sequoia* (*seq*) (Gunnar et al. 2016), which encodes a C2H2 zinc-finger transcription factor homologous to *Drosophila* Tramtrack. In addition, the screen identified *Ctr9* (*Cln Three Requiring 9*) (Bahrampour and Thor 2016), which encodes a key component of the Paf1 epigenetic complex. Both *seq* and *Ctr9* affect the expression of Notch pathway components, but their interplay with Notch signalling is nebulous (Bahrampour and Thor 2016; Gunnar et al. 2016).

Dictating Asymmetric Cell Fate Between Sibling Neurons

The third role for Notch signalling pertains to its role in controlling asymmetric cell division of GMCs. Intriguingly, not only is the NB->GMC division asymmetric (as outlined above), in most, if not all, cases when a GMC divides, it also divides asymmetrically. While Notch signalling does not appear to control the asymmetric NB->GMC division, it does influence the GMC->2-neurons asymmetric division.

Specifically, as each GMC divides, it generates two daughter cells that differentiate, which usually, perhaps always, acquire different cell fate. The most well-studied examples of this are the aCC/pCC, RP2/RP2sib, U/Usib and dMP2/vMP2 sibling pairs, where the two siblings are easily distinguishable by both markers and axonal projections (Garces and Thor 2006; Bhat et al. 2011; Skeath and Doe 1998; Spana and Doe 1996). Studies of these model sibling pairs have revealed that a number of cell fate determinants distribute unequally between the two daughter cells and act therein to govern different cell fates, by regulating Notch signalling. One key such asymmetric determinant is the Numb protein (Doe 1996; Fuerstenberg et al. 1998; Doe and Bowerman 2001; Spana and Doe 1996). Numb segregates into one daughter cell, where it inhibits Notch signalling dictating the default "Notch OFF", or "B" fate. By contrast, the absence of Numb in the other sibling cell allows for Notch signalling, and this cell then acquires the "Notch ON" or "A" fate (Cau and Blader 2009). In addition, the transmembrane protein Sanpodo (Spdo) also regulates Notch signalling only during the asymmetric division decision and promotes the "Notch ON" fate (Skeath and Doe 1998; O'Connor-Giles and Skeath 2003) (Fig. 3.5). Spdo plays two roles upon the Notch pathway, acting both to amplify Notch signalling in the Numb-lacking cell and to inhibit Notch signalling in the Numb-expressing cell (Babaoglan et al. 2009). The precise underlying molecular mechanisms of Spdo function upon Notch are still nebulous. Moreover, why Spdo only acts to modify Notch signalling during GMC asymmetric division is also unclear. Intriguingly, the activation of Notch signalling in the "A" neuron, as an effect of "A" cells having no Numb protein, results in the activation of a different HES-related TF, Hey, which is not activated by Notch in the early ectoderm (Monastirioti et al. 2010; Ulvklo et al. 2012). The role of *hey* is still unclear, although mutants are late embryonic/early larval lethal, and there was no apparent effect upon the sibling decision in several of the model pairs. However, misexpression of *hey* could indeed affect sibling cell fate, imposing "A" fate in "B" siblings (Monastirioti et al. 2010). Understanding Notch activation of the *E(spl)* HES genes versus *hey* may provide a powerful means of addressing context dependency of Notch nuclear output.

In many cases, it has been found that asymmetric Notch signalling not only dictates two different postmitotic neuronal cell fates, but it can also gate between programmed cell death (PCD) and survival (Lundell et al. 2003; Miguel-Aliaga and Thor 2004; Bhat et al. 2011). Notch signalling can act in either direction and allow for the survival, and therefore the specification of a particular neural fate, or can induce PCD. Global mapping studies reveal that most, if not all, NBs produce cells that become subsequently removed by PCD (Rogulja-Ortmann et al. 2007). If PCD

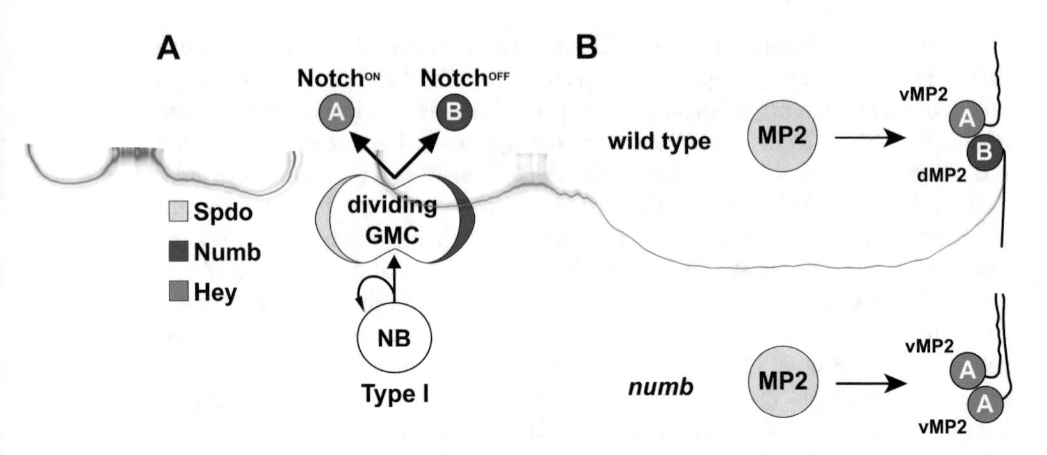

Fig. 3.5 Notch controls asymmetric sibling cell fate. (**a**) Schematic presentation of a dividing GMC, born within a Type I window. GMCs divide asymmetrically and segregate Spdo and Numb to opposite poles. After GMC cell division, the sibling cell that inherits Numb protein will be Notch-OFF and acquire fate B, while the sibling cell inheriting Spdo protein will be Notch-ON, express the Hey bHLH factor and acquire fate A. (**b**) A well-studied example of a Notch-mediated sibling decision is the vMP2/dMP2 sibling pairs. In wild type, the MP2 progenitor cell undergoes an asymmetric cell division, generating two different types of neurons, projecting their axons in opposite directions. In *numb* mutants, both neurons become Notch-ON and acquire the vMP2 cell fate

is blocked, cells that normally are removed by PCD can often differentiate as neurons and project axons. The stereotyped PCD observed in many siblings in the *Drosophila* developing CNS can be likened to the specification of a unique cell fate within the nervous system (Miguel-Aliaga and Thor 2009).

Glia Development

The fourth role of the Notch pathway during *Drosophila* embryonic CNS development pertains to the role of the Notch receptor in glia development. This aspect of Notch pathway function stems from findings that the longitudinal axon scaffold is lost in Notch mutants (Giniger 1998; Giniger et al. 1993). Longitudinal axons form on each side of the VNC midline and contain a large number of interneuron axons, projecting up and down the VNC. These tracts develop in intimate relation to a set of specialized glia, the longitudinal glia (LG). By using a number of selective genetic tools, such as temperature-sensitive alleles and Gal4/UAS-driven transgenic expression using highly selective drivers, a picture has emerged where it appears that Dl is provided by longitudinal axons, which activates canonical Notch signalling in the longitudinal glia (LG). Activation of Notch in LG is promoted by Fringe,

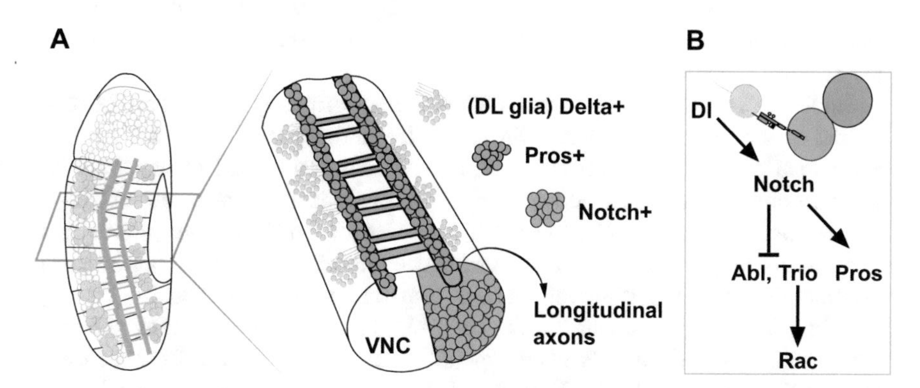

Fig. 3.6 Notch controls glia development. (**a**) Cartoon depicting *Drosophila* embryo on the left and a magnified section view of the VNC on the right, where longitudinal interneuron axons project on each side of the midline, supported by longitudinal glia (LG). (**b**) The longitudinal axons present Dl to the longitudinal glia (LG) and activate canonical Notch signalling, which activates Pros expression and represses Abl, Tri and Rac signalling

a glycosylation enzyme that enhances Notch responsiveness to the Dl ligand (see 2 above). Activated Notch, in turn, activates Pros, and Pros promotes LG fate in several ways, including promoting proliferation and differentiation (Thomas and van Meyel 2007; Griffiths et al. 2007; Griffiths and Hidalgo 2004) (Fig. 3.6). A key effector gene downstream of Notch-Fng signalling is the glutamate transporter *Eaat1*, a key enzyme for LG function (Stacey et al. 2010). Complicating the issue is that non-canonical Notch signalling also acts inside axons that project along longitudinal connectives, which are ensheathed by LGs (Kuzina et al. 2011).

Axon Pathfinding

The fifth role of the Notch pathway during *Drosophila* embryonic CNS development pertains to the role of the Notch receptor in motor axon pathfinding.

The *Drosophila* neuromuscular system has been a useful experimental framework for studies aimed at addressing the mechanisms underlying the assembly of neuronal networks. In the *Drosophila* embryonic nerve cord, in each abdominal hemisegment, a stereotyped array of ~40 motor neurons innervate a stereotyped scaffold of 30 somatic muscles (Landgraf et al. 1997; Schmid et al. 1999). The axons from the ~40 motor neurons project out of the nerve cord along three main branches, the intersegmental nerve (ISN), the segmental nerve (SN) and the transverse nerve (TN). Two of these main branches, the ISN and SN, branch further into sub-branches: the ISN, ISNb, ISNd, SNa and SNc. The ISN class of motor neurons project their axons dorsally, while those of the other five classes project laterally and ventrally. Hence, based on their axonal projections, the ~40 motor neurons can be subdivided into six distinct subclasses and display a general 1:1, muscle/motor neuron, connection ratio [reviewed in Landgraf and Thor (2006) and Thor and Thomas (2002)].

Studies of Notch in motor axon pathfinding have primarily focused on the pathfinding of the ISNb motor nerve. These studies initially relied on a temperature-sensitive allele of Notch (*Notch*^{ts1}), which allowed for removing Notch activity late in embryonic development, well after most neuronal identities have been specified. Late temperature shift resulted in ISNb motor axons bypassing their normal exit point, where they typically enter into their target muscle field (Fig. 3.7). Moreover, a similar phenotype was observed for a *Dl* temperature-sensitive allele. Importantly, providing *Dl* activity to tracheal cells, i.e. back to the cells that constitute the axon choice point, rescued the *Dl* ts phenotype. In contrast, there was no effect in *Serrate* mutants (Crowner et al. 2003; Giniger et al. 1993). Further studies revealed that Abl interacted genetically with Notch, but not Su(H), and that overexpression of Notch suppressed the Abl overexpression phenotype. Hence, Notch acts in a non-canonical manner with respect to motor axon pathfinding, via interaction with the Abl tyrosine kinase signalling network (Crowner et al. 2003; Giniger 1998; Kuzina et al. 2011; Le Gall et al. 2008) (see section "Notch Non-canonical Molecular Pathway" for more detailed description of the molecular details). Briefly, a combination of biochemical, molecular and genetic studies have demonstrated that upon activation by Delta, Notch promotes the growth and guidance of motor axons in the *Drosophila* embryo by locally suppressing the Abl signalling events (Crowner et al. 2003; Giniger 1998; Kuzina et al. 2011).

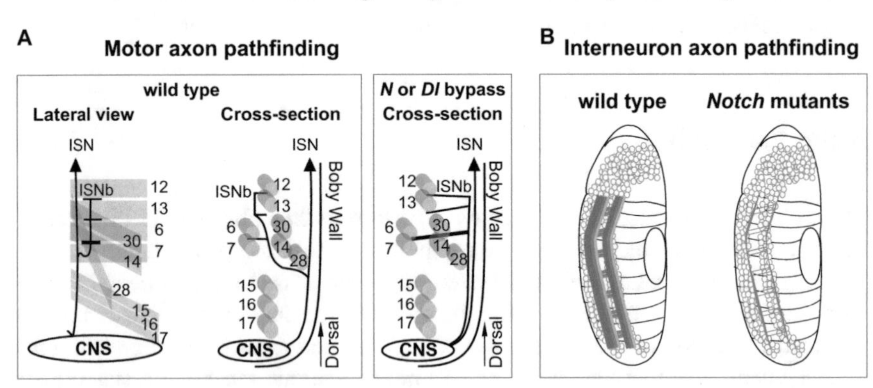

Fig. 3.7 Notch controls axon pathfinding. Schematic illustration of non-canonical Notch pathway regulation of motor axon pathfinding during *Drosophila* embryonic CNS development. (**a**) Motor axons project along the two sub-branches of the intersegmental (ISN) motor neurons: ISN and ISNb. The ISN motor neurons project axons dorsally, while ISNb motor axons leave the main ISN fascicle to innervate the lateral muscles. In *Notch* or *Dl* mutants, ISNb axons bypass their exit point and continue along the common ISN tract. (**b**) Schematic illustration of interneuron axons in a wild-type and *Notch* mutant *Drosophila* embryo. (See the text for more detailed description)

Conclusions

Studies of the developing *Drosophila* CNS, and in particular the VNC, has unravelled a daunting complexity of how the Notch signal transduction pathway is used, demonstrating that Notch signalling acts during, at least, five different stages, controlling five entirely different biological outcomes (Fig. 3.8). The five events, i.e. NB selection, Type I->0 daughter proliferation switch, asymmetric GMC division, glia proliferation/fate and motor axon pathfinding, occur during a short time span of embryonic development.

The use of conditional mutants (temperature-sensitive alleles) and highly selective transgenic expression of a range of dominant-negative and dominant-activated constructs, as well as the use of selective reporters and markers of Notch activity, has allowed investigators to dissect these different roles of Notch from one another.

A single NB may be involved in at least four out of five Notch-mediated decisions. While our current understanding of Notch signalling reveals that some of this gating is controlled by asymmetric distribution of a number of different proteins, it is still intriguing to ponder the dynamic control of Notch signalling.

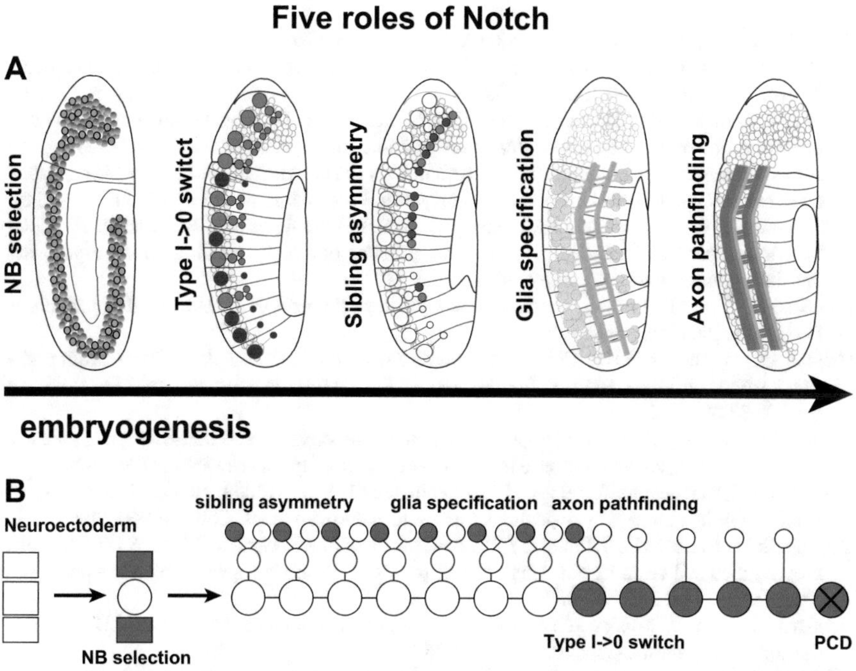

Fig. 3.8 Summary of the five events of *Drosophila* embryonic CNS development where Notch plays a key role. Carton highlighting the five aspects of neurodevelopment that are gated by Notch signalling. (See the text for more detailed description)

There are a number of outstanding issues regarding Notch signalling in the *Drosophila* embryonic CNS: How is the context dependency of Notch controlled, other TFs, co-TFs, chromatin? For example, how is activation of E(spl) in NBs versus Hey in siblings controlled? Why do only certain components, such as Spdo, act in certain decisions? How is the choice between canonical and non-canonical Notch signalling controlled?

In mammals, Notch signalling plays a number of roles during CNS development. However, given that novel roles for Notch signalling have emerged recently in the high-resolution and genetically tractable *Drosophila* system, it is likely that the study of Notch biology will yield many surprising findings in the years to come also in the mammalian CNS.

Author Contributions SB and ST generated the figures and wrote the manuscript.

Competing Interests No competing interests declared.

Funding Funding was provided by the Swedish Research Council (2018-00694) to SB and by support from the University of Queensland to ST.

References

Acar M, Jafar-Nejad H, Takeuchi H, Rajan A, Ibrani D, Rana NA, Pan H, Haltiwanger RS, Bellen HJ (2008) Rumi is a CAP10 domain glycosyltransferase that modifies Notch and is required for Notch signaling. Cell 132:247–258

Allan DW, Thor S (2015) Transcriptional selectors, masters, and combinatorial codes: regulatory principles of neural subtype specification. Wiley Interdiscip Rev Dev Biol 4:505–528

Artavanis-Tsakonas S, Muskavitch MA, Yedvobnick B (1983) Molecular cloning of Notch, a locus affecting neurogenesis in Drosophila melanogaster. Proc Natl Acad Sci U S A 80:1977–1981

Babaoglan AB, O'Connor-Giles KM, Mistry H, Schickedanz A, Wilson BA, Skeath JB (2009) Sanpodo: a context-dependent activator and inhibitor of Notch signaling during asymmetric divisions. Development 136:4089–4098

Bachmann A, Knust E (1998) Dissection of cis-regulatory elements of the Drosophila gene Serrate. Dev Genes Evol 208:346–351

Bahrampour S, Thor S (2016) Ctr9, a key component of the Paf1 complex, affects proliferation and terminal differentiation in the developing Drosophila nervous system. G3 (Bethesda) 6:3229–3239

Bailey AM, Posakony JW (1995) Suppressor of hairless directly activates transcription of enhancer of split complex genes in response to Notch receptor activity. Genes Dev 9:2609–2622

Baumgardt M, Karlsson D, Terriente J, Diaz-Benjumea FJ, Thor S (2009) Neuronal subtype specification within a lineage by opposing temporal feed-forward loops. Cell 139:969–982

Baumgardt M, Karlsson D, Salmani BY, Bivik C, Macdonald RB, Gunnar E, Thor S (2014) Global programmed switch in neural daughter cell proliferation mode triggered by a temporal gene cascade. Dev Cell 30:192–208

Bender LB, Kooh PJ, Muskavitch MA (1993) Complex function and expression of Delta during Drosophila oogenesis. Genetics 133:967–978

Bhat KM (1999) Segment polarity genes in neuroblast formation and identity specification during Drosophila neurogenesis. Bioessays 21:472–485

Bhat KM, Gaziova I, Katipalla S (2011) Neuralized mediates asymmetric division of neural precursors by two distinct and sequential events: promoting asymmetric localization of Numb and enhancing activation of Notch-signaling. Dev Biol 351:186–198

Birkholz O, Rickert C, Berger C, Urbach R, Technau GM (2013) Neuroblast pattern and identity in the Drosophila tail region and role of double sex in the survival of sex-specific precursors. Development 140:1830–1842

Bivik Stadler C, Arefin B, Ekman H, Thor S (2019) PIP degron-stabilized Dacapo/p21(Cip1) and mutations in ago act in an anti- versus pro-proliferative manner, yet both trigger an increase in Cyclin E levels. Development 146(13):dev175927

Bivik C, Macdonald RB, Gunnar E, Mazouni K, Schweisguth F, Thor S (2016) Control of neural daughter cell proliferation by multi-level Notch/Su(H)/E(spl)-HLH signaling. PLoS Genet 12:e1005984

Boone JQ, Doe CQ (2008) Identification of Drosophila type II neuroblast lineages containing transit amplifying ganglion mother cells. Dev Neurobiol 68:1185–1195

Bray SJ (2006) Notch signalling: a simple pathway becomes complex. Nat Rev Mol Cell Biol 7:678–689

Cau E, Blader P (2009) Notch activity in the nervous system: to switch or not switch? Neural Dev 4:36

Choksi SP, Southall TD, Bossing T, Edoff K, de Wit E, Fischer BE, van Steensel B, Micklem G, Brand AH (2006) Prospero acts as a binary switch between self-renewal and differentiation in Drosophila neural stem cells. Dev Cell 11:775–789

Cohen SM, Jurgens G (1990) Mediation of Drosophila head development by gap-like segmentation genes. Nature 346:482–485

Crowner D, Le Gall M, Gates MA, Giniger E (2003) Notch steers Drosophila ISNb motor axons by regulating the Abl signaling pathway. Curr Biol 13:967–972

Deblandre GA, Lai EC, Kintner C (2001) Xenopus neuralized is a ubiquitin ligase that interacts with XDelta1 and regulates Notch signaling. Dev Cell 1:795–806

Delidakis C, Preiss A, Hartley DA, Artavanistsakonas S (1991) 2 genetically and molecularly distinct functions involved in early neurogenesis reside within the enhancer of split locus of Drosophila-melanogaster. Genetics 129:803–823

Doe CQ (1996) Asymmetric cell division and neurogenesis. Curr Opin Genet Dev 6:562–566

Doe CQ, Bowerman B (2001) Asymmetric cell division: fly neuroblast meets worm zygote. Curr Opin Cell Biol 13:68–75

Doe CQ, Technau GM (1993) Identification and cell lineage of individual neural precursors in the Drosophila CNS. Trends Neurosci 16:510–514

Fleming RJ, Scottgale TN, Diederich RJ, Artavanis-Tsakonas S (1990) The gene Serrate encodes a putative EGF-like transmembrane protein essential for proper ectodermal development in Drosophila melanogaster. Genes Dev 4:2188–2201

Fortini ME (2009) Notch signaling: the core pathway and its posttranslational regulation. Dev Cell 16:633–647

Fortini ME, Artavanis-Tsakonas S (1994) The suppressor of hairless protein participates in notch receptor signaling. Cell 79:273–282

Fuerstenberg S, Broadus J, Doe CQ (1998) Asymmetry and cell fate in the Drosophila embryonic CNS. Int J Dev Biol 42:379–383

Garces A, Thor S (2006) Specification of Drosophila aCC motoneuron identity by a genetic cascade involving even-skipped, grain and zfh1. Development 133:1445–1455

Giniger E (1998) A role for Abl in Notch signaling. Neuron 20:667–681

Giniger E (2012) Notch signaling and neural connectivity. Curr Opin Genet Dev 22:339–346

Giniger E, Jan LY, Jan YN (1993) Specifying the path of the intersegmental nerve of the Drosophila embryo: a role for Delta and Notch. Development 117:431–440

Griffiths RL, Hidalgo A (2004) Prospero maintains the mitotic potential of glial precursors enabling them to respond to neurons. EMBO J 23:2440–2450

Griffiths RC, Benito-Sipos J, Fenton JC, Torroja L, Hidalgo A (2007) Two distinct mechanisms segregate Prospero in the longitudinal glia underlying the timing of interactions with axons. Neuron Glia Biol 3:75–88

Gunnar E, Bivik C, Starkenberg A, Thor S (2016) sequoia controls the type I>0 daughter proliferation switch in the developing Drosophila nervous system. Development 143:3774–3784

Handford PA, Korona B, Suckling R, Redfield C, Lea SM (2018) Structural insights into Notch receptor-ligand interactions. Adv Exp Med Biol 1066:33–46

Hirata J, Nakagoshi H, Nabeshima Y, Matsuzaki F (1995) Asymmetric segregation of the homeodomain protein Prospero during Drosophila development. Nature 377:627–630

Hirth F, Reichert H (1999) Conserved genetic programs in insect and mammalian brain development. Bioessays 21:677–684

Jennings B, Preiss A, Delidakis C, Bray S (1994) The Notch signalling pathway is required for Enhancer of split bHLH protein expression during neurogenesis in the Drosophila embryo. Development 120:3537–3548

Johansen KM, Fehon RG, Artavanis-Tsakonas S (1989) The notch gene product is a glycoprotein expressed on the cell surface of both epidermal and neuronal precursor cells during Drosophila development. J Cell Biol 109:2427–2440

Kannan R, Song JK, Karpova T, Clarke A, Shivalkar M, Wang B, Kotlyanskaya L, Kuzina I, Gu Q, Giniger E (2017) The Abl pathway bifurcates to balance Enabled and Rac signaling in axon patterning in Drosophila. Development 144:487–498

Kannan R, Cox E, Wang L, Kuzina I, Gu Q, Giniger E (2018) Tyrosine phosphorylation and proteolytic cleavage of Notch are required for non-canonical Notch/Abl signaling in Drosophila axon guidance. Development 145(2):dev151548

Kidd S, Baylies MK, Gasic GP, Young MW (1989) Structure and distribution of the Notch protein in developing Drosophila. Genes Dev 3:1113–1129

Klambt C, Knust E, Tietze K, Campos-Ortega JA (1989) Closely related transcripts encoded by the neurogenic gene complex Enhancer of split of *Drosophila melanogaster*. EMBO J 8:203–210

Knoblich JA, Jan LY, Jan YN (1995) Asymmetric segregation of Numb and Prospero during cell division. Nature 377:624–627

Knust E, Schrons H, Grawe F, Campos-Ortega JA (1992) Seven genes of the Enhancer of split complex of Drosophila melanogaster encode helix-loop-helix proteins. Genetics 132:505–518

Kopczynski CC, Muskavitch MA (1989) Complex spatio-temporal accumulation of alternative transcripts from the neurogenic gene Delta during Drosophila embryogenesis. Development 107:623–636

Kuzina I, Song JK, Giniger E (2011) How Notch establishes longitudinal axon connections between successive segments of the Drosophila CNS. Development 138:1839–1849

Lai EC (2002) Protein degradation: four E3s for the notch pathway. Curr Biol 12:R74–R78

Lai EC, Deblandre GA, Kintner C, Rubin GM (2001) Drosophila neuralized is a ubiquitin ligase that promotes the internalization and degradation of delta. Dev Cell 1:783–794

Landgraf M, Thor S (2006) Development of Drosophila motoneurons: specification and morphology. Semin Cell Dev Biol 17:3–11

Landgraf M, Bossing T, Technau GM, Bate M (1997) The origin, location, and projections of the embryonic abdominal motorneurons of Drosophila. J Neurosci 17:9642–9655

Lawrence PA, Sanson B, Vincent JP (1996) Compartments, wingless and engrailed: patterning the ventral epidermis of Drosophila embryos. Development 122:4095–4103

Le Gall M, de Mattei C, Giniger E (2008) Molecular separation of two signaling pathways for the receptor, Notch. Dev Biol 313:556–567

Lecourtois M, Schweisguth F (1995) The neurogenic suppressor of hairless DNA-binding protein mediates the transcriptional activation of the enhancer of split complex genes triggered by Notch signaling. Genes Dev 9:2598–2608

Li L, Vaessin H (2000) Pan-neural Prospero terminates cell proliferation during Drosophila neurogenesis. Genes Dev 14:147–151

Lieber T, Kidd S, Young MW (2002) kuzbanian-mediated cleavage of Drosophila Notch. Genes Dev 16:209–221

Lundell MJ, Lee HK, Perez E, Chadwell L (2003) The regulation of apoptosis by Numb/Notch signaling in the serotonin lineage of Drosophila. Development 130:4109–4121

Miguel-Aliaga I, Thor S (2004) Segment-specific prevention of pioneer neuron apoptosis by cell-autonomous, postmitotic Hox gene activity. Development 131:6093–6105

Miguel-Aliaga I, Thor S (2009) Programmed cell death in the nervous system–a programmed cell fate? Curr Opin Neurobiol 19:127–133

Miller AC, Lyons EL, Herman TG (2009) cis-Inhibition of Notch by endogenous Delta biases the outcome of lateral inhibition. Curr Biol 19:1378–1383

Moloney DJ, Panin VM, Johnston SH, Chen J, Shao L, Wilson R, Wang Y, Stanley P, Irvine KD, Haltiwanger RS, Vogt TF (2000) Fringe is a glycosyltransferase that modifies Notch. Nature 406:369–375

Monastirioti M, Giagtzoglou N, Koumbanakis KA, Zacharioudaki E, Deligiannaki M, Wech I, Almeida M, Preiss A, Bray S, Delidakis C (2010) Drosophila Hey is a target of Notch in asymmetric divisions during embryonic and larval neurogenesis. Development 137:191–201

Nakao K, Campos-Ortega JA (1996) Persistent expression of genes of the enhancer of split complex suppresses neural development in Drosophila. Neuron 16:275–286

Nam Y, Sliz P, Song L, Aster JC, Blacklow SC (2006) Structural basis for cooperativity in recruitment of MAML coactivators to Notch transcription complexes. Cell 124:973–983

Newsome TP, Schmidt S, Dietzl G, Keleman K, Asling B, Debant A, Dickson BJ (2000) Trio combines with dock to regulate Pak activity during photoreceptor axon pathfinding in Drosophila. Cell 101:283–294

Nicholson SC, Nicolay BN, Frolov MV, Moberg KH (2011) Notch-dependent expression of the archipelago ubiquitin ligase subunit in the Drosophila eye. Development 138:251–260

O'Connor-Giles KM, Skeath JB (2003) Numb inhibits membrane localization of Sanpodo, a four-pass transmembrane protein, to promote asymmetric divisions in Drosophila. Dev Cell 5:231–243

Oellers N, Dehio M, Knust E (1994) bHLH proteins encoded by the Enhancer of split complex of Drosophila negatively interfere with transcriptional activation mediated by proneural genes. Mol Gen Genet 244:465–473

Okajima T, Xu A, Irvine KD (2003) Modulation of notch-ligand binding by protein O-fucosyltransferase 1 and fringe. J Biol Chem 278:42340–42345

Pan D, Rubin GM (1997) Kuzbanian controls proteolytic processing of Notch and mediates lateral inhibition during Drosophila and vertebrate neurogenesis. Cell 90:271–280

Panin VM, Papayannopoulos V, Wilson R, Irvine KD (1997) Fringe modulates Notch-ligand interactions. Nature 387:908–912

Pavlopoulos E, Pitsouli C, Klueg KM, Muskavitch MA, Moschonas NK, Delidakis C (2001) neuralized Encodes a peripheral membrane protein involved in delta signaling and endocytosis. Dev Cell 1:807–816

Petcherski AG, Kimble J (2000) Mastermind is a putative activator for Notch. Curr Biol 10:R471–R473

Pitsouli C, Delidakis C (2005) The interplay between DSL proteins and ubiquitin ligases in Notch signaling. Development 132:4041–4050

Poulson DF (1937) Chromosomal deficiencies and the embryonic development of Drosophila Melanogaster. Proc Natl Acad Sci U S A 23:133–137

Qi H, Rand MD, Wu X, Sestan N, Wang W, Rakic P, Xu T, Artavanis-Tsakonas S (1999) Processing of the notch ligand delta by the metalloprotease Kuzbanian. Science 283:91–94

Rogulja-Ortmann A, Luer K, Seibert J, Rickert C, Technau GM (2007) Programmed cell death in the embryonic central nervous system of Drosophila melanogaster. Development 134:105–116

Schmid A, Chiba A, Doe CQ (1999) Clonal analysis of Drosophila embryonic neuroblasts: neural cell types, axon projections and muscle targets. Development 126:4653–4689

Schweisguth F, Posakony JW (1994) Antagonistic activities of Suppressor of Hairless and Hairless control alternative cell fates in the Drosophila adult epidermis. Development 120:1433–1441

Simoes S, Oh Y, Wang MFZ, Fernandez-Gonzalez R, Tepass U (2017) Myosin II promotes the anisotropic loss of the apical domain during Drosophila neuroblast ingression. J Cell Biol 216:1387–1404

Skeath JB (1999) At the nexus between pattern formation and cell-type specification: the generation of individual neuroblast fates in the Drosophila embryonic central nervous system. Bioessays 21:922–931

Skeath JB, Doe CQ (1998) Sanpodo and Notch act in opposition to Numb to distinguish sibling neuron fates in the Drosophila CNS. Development 125:1857–1865

Skeath JB, Thor S (2003) Genetic control of Drosophila nerve cord development. Curr Opin Neurobiol 13:8–15

Song JK, Giniger E (2011) Noncanonical Notch function in motor axon guidance is mediated by Rac GTPase and the GEF1 domain of Trio. Dev Dyn 240:324–332

Spana EP, Doe CQ (1995) The prospero transcription factor is asymmetrically localized to the cell cortex during neuroblast mitosis in Drosophila. Development 121:3187–3195

Spana EP, Doe CQ (1996) Numb antagonizes Notch signaling to specify sibling neuron cell fates. Neuron 17:21–26

Stacey SM, Muraro NI, Peco E, Labbe A, Thomas GB, Baines RA, van Meyel DJ (2010) Drosophila glial glutamate transporter Eaat1 is regulated by fringe-mediated notch signaling and is essential for larval locomotion. J Neurosci 30:14446–14457

Thomas GB, van Meyel DJ (2007) The glycosyltransferase Fringe promotes Delta-Notch signaling between neurons and glia, and is required for subtype-specific glial gene expression. Development 134:591–600

Thomas U, Speicher SA, Knust E (1991) The Drosophila gene Serrate encodes an EGF-like transmembrane protein with a complex expression pattern in embryos and wing discs. Development 111:749–761

Thor S (1995) The genetics of brain-development – conserved programs in flies and mice. Neuron 15:975–977

Thor S, Thomas J (2002) Motor neuron specification in worms, flies and mice: conserved and 'lost' mechanisms. Curr Opin Genet Dev 12:558–564

Ulvklo C, Macdonald R, Bivik C, Baumgardt M, Karlsson D, Thor S (2012) Control of neuronal cell fate and number by integration of distinct daughter cell proliferation modes with temporal progression. Development 139:678–689

Urbach R, Schnabel R, Technau GM (2003) The pattern of neuroblast formation, mitotic domains and proneural gene expression during early brain development in Drosophila. Development 130:3589–3606

Urbach R, Jussen D, Technau GM (2016) Gene expression profiles uncover individual identities of gnathal neuroblasts and serial homologies in the embryonic CNS of Drosophila. Development 143:1290–1301

Varshney S, Stanley P (2018) Multiple roles for O-glycans in Notch signalling. FEBS Lett 592:3819–3834

Vassin H, Bremer KA, Knust E, Campos-Ortega JA (1987) The neurogenic gene Delta of Drosophila melanogaster is expressed in neurogenic territories and encodes a putative transmembrane protein with EGF-like repeats. EMBO J 6:3431–3440

Wang Y, Shao L, Shi S, Harris RJ, Spellman MW, Stanley P, Haltiwanger RS (2001) Modification of epidermal growth factor-like repeats with O-fucose. Molecular cloning and expression of a novel GDP-fucose protein O-fucosyltransferase. J Biol Chem 276:40338–40345

Wharton KA, Johansen KM, Xu T, Artavanis-Tsakonas S (1985) Nucleotide sequence from the neurogenic locus Notch implies a gene product that shares homology with proteins containing EGF-like repeats. Cell 43:567–581

Wheeler SR, Stagg SB, Crews ST (2009) MidExDB: a database of Drosophila CNS midline cell gene expression. BMC Dev Biol 9:56

Wilson JJ, Kovall RA (2006) Crystal structure of the CSL-Notch-Mastermind ternary complex bound to DNA. Cell 124:985–996

Wu L, Aster JC, Blacklow SC, Lake R, Artavanis-Tsakonas S, Griffin JD (2000) MAML1, a human homologue of Drosophila mastermind, is a transcriptional co-activator for NOTCH receptors. Nat Genet 26:484–489

Yeh CH, Bellon M, Nicot C (2018) FBXW7: a critical tumor suppressor of human cancers. Mol Cancer 17:115

Zweifel ME, Leahy DJ, Hughson FM, Barrick D (2003) Structure and stability of the ankyrin domain of the Drosophila Notch receptor. Protein Sci 12:2622–2632

Chapter 4
Epigenetic Regulation of Notch Signaling During *Drosophila* Development

Chuanxian Wei, Chung-Weng Phang, and Renjie Jiao

Abstract Notch signaling exerts multiple important functions in various developmental processes, including cell differentiation and cell proliferation, while misregulation of this pathway results in a variety of complex diseases, such as cancer and developmental defects. The simplicity of the Notch pathway in *Drosophila melanogaster*, in combination with the availability of powerful genetics, makes this an attractive model for studying the fundamental mechanisms of how Notch signaling is regulated and how it functions in various cellular contexts. Recently, increasing evidence for epigenetic control of Notch signaling reveals the intimate link between epigenetic regulators and Notch signaling pathway. In this chapter, we summarize the research advances of Notch and CAF-1 in *Drosophila* development and the epigenetic regulation mechanisms of Notch signaling activity by CAF-1 as well as other epigenetic modification machineries, which enables Notch to orchestrate different biological inputs and outputs in specific cellular contexts.

Keywords Notch · CAF-1 · Signal transduction · Gene expression · Epigenetic regulation · Chromatin assembly factors · Development · *Drosophila*

Introduction

The Notch pathway is an evolutionarily conserved signaling cascade present in most of multicellular organisms and plays important roles in development and physiology. Notch signaling regulates a variety of biological processes, including cell proliferation, differentiation, and apoptosis (Radtke et al. 2005). Mis-regulation of Notch signaling activity has been associated with various complex diseases, such as

C. Wei · C.-W. Phang · R. Jiao (✉)
Sino-French Hoffmann Institute, School of Basic Medical Sciences, Guangzhou Medical University, Guangzhou, China
e-mail: rjiao@gzhmu.edu.cn

J. Reichrath, S. Reichrath (eds.), *Notch Signaling in Embryology and Cancer*, Advances in Experimental Medicine and Biology 1218,
https://doi.org/10.1007/978-3-030-34436-8_4

cancer and neurological disorders (Salazar and Yamamoto 2018). Understanding the mechanisms of signaling regulation, as well as the outcome of signaling in various tissues, is therefore of great importance. The existence of multiple paralogues of Notch receptor (Notch 1–4) and ligands (Delta 1–4 and Jagged 1–2) in mammals and other vertebrates makes Notch-related studies more complicated in those animals. However, the situation is much simpler in *Drosophila*, which has only one Notch receptor and two ligands, Delta (Dl) and Serrate (Ser). All of these three proteins share highly conserved sequences with their mammalian counterparts (Muskavitch 1994). The simplicity of the Notch signaling pathway in *Drosophila*, in combination with the availability of well-established powerful genetic tools and materials (Zacharioudaki and Bray 2014), makes *Drosophila* an extremely attractive system for studying Notch pathway. Recent studies in both *Drosophila* and mammals provide insights into the epigenetic regulation of Notch signaling, and this chapter summarizes the current understanding of how Notch signaling is epigenetically regulated, mainly by CAF-1.

Notch Signaling

A century ago, Notch was first described as a wing margin developmental defect phenotype in *Drosophila melanogaster* (Bray 2016; Ntziachristos et al. 2014). Notch locus was identified as a gene that is responsible for the notched wing phenotype (Welshons 1958a, b), which gives the name to the pathway. Notch gene encodes a single-pass type I transmembrane receptor, the extracellular domain of which includes a variable number of EGF-like repeats, with the functions of ligand binding (Fig. 4.1).

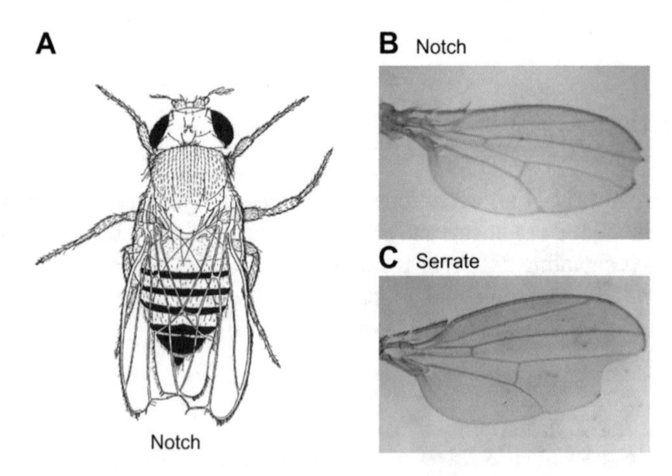

Fig. 4.1 Phenotype of *Drosophila* with Notch pathway mutations. (**a**) Drawing of a Notch receptor mutant fly with a notched wing tip. (**b**, **c**) Photo of a wing from a fly carrying a Notch mutation (**b**) and a mutation in Serrate (**c**). (Images are adapted from Alabi et al. (2018))

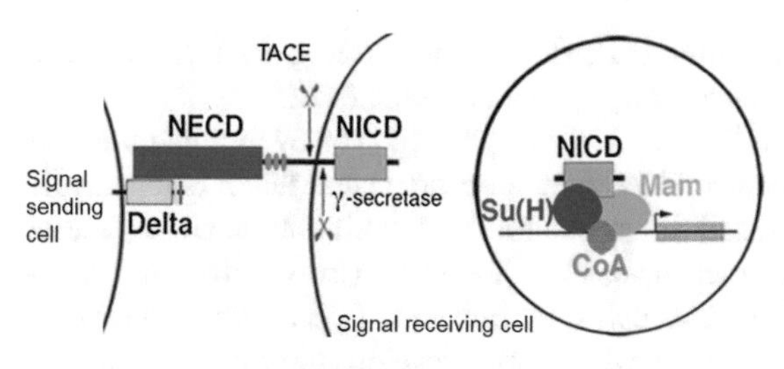

	Drosophila	Mammals
Ligand	Delta, Serrate	Delta 1-4, Jagged 1-2
Receptor	Notch	Notch 1-4
First cleavage	Kuzbanian, TACE	ADAM10, ADAM17
γ-secretase	Presenilin, Nicastrin, PEN2 and APH1	Presenilin1-2, Nicastrin, PEN2 and APH1
Effector	Notch intracellular domain (NICD)	Notch intracellular domain (NICD)
Transcriptional factor	Su(H)	CBF-1
Co-activator	Mastermind(Mam)	Mastermind(Mam) 1-3

Fig. 4.2 The core Notch signaling pathway and the main components of Notch pathway in *Drosophila* and mammals

Mutants with defects in other genes that are part of the Notch pathway were later identified because they had similar phenotypes and were named Delta (Dl) and Serrate (Ser) (Siren and Portin 1989; Shepard et al. 1989; Thomas et al. 1991). Dl and Ser are also transmembrane proteins and share the similar EGF repeats with Notch. Thus, in order for Notch signaling to occur, the ligand-expressing cells (or signal sending cells) have to be in intimate contact with the receptor-expressing cells (or signal receiving cells) (Fig. 4.2).

The canonical Notch signaling pathway is rather simple. While vertebrates have several Notch receptors and ligands, the *Drosophila* genome only contains one Notch receptor and two ligands, Delta (Dl) and Serrate (Ser) (Klueg and Muskavitch 1999). Like many other signaling pathways, Notch signaling is initiated by receptor-ligand interaction between neighboring cells with close contact or direct contact. The Notch receptor can be activated by binding with the ligands Dl or Ser that are expressed in adjacent cells. Upon activation by this intimate binding, the Notch receptor undergoes two consecutive cleavage events, which are catalyzed sequentially by an ADAM family metalloprotease (Kuz and Tace in *Drosophila*) (Alabi et al. 2018) and by the γ-secretase complex (containing Presenilin/Psn, Nicastrin/Nct, PEN2, and APH1) (De Strooper et al. 1999; Yang et al. 2019), resulting in the release of the intracellular portion of the protein, called the Notch intracellular domain (NICD), which then migrates into the nucleus and joins a protein complex directly bound to chromatin to initiate the transcription of target genes (Borggrefe and Oswald 2016). This complex includes the transcription factors Suppressor of Hairless (Su(H)), as well as other potential co-regulators, such as the transcriptional coactivator Mastermind (Mam), thereby leading to the expression of Notch-dependent target genes (Borggrefe and Liefke 2012).

As we mentioned above, Notch signaling is an evolutionally conserved pathway throughout metazoans. Over time, it became clear that it is repeatedly employed in cell fate decisions, cell differentiation, cell proliferation, and cell survival in diverse contexts and at distinct stages of development (Bray 2016). In many developmental contexts, Notch specifies cell fate decisions. In the developing vertebrate eye, for example, Notch regulates which cells develop into glial cells and which develop into optic neurons (Genethliou et al. 2009). Not surprisingly, mutations leading to dysregulated Notch signaling have also been implicated in cancer, including hematological malignancies (Bugeon et al. 2011) and solid tumors (Mutvei et al. 2015). Mis-regulation of Notch signaling in ovarian follicle cells disturbs the balance between cell proliferation and cell differentiation in *Drosophila* oogenesis, leading to cell death and sterility (Deng et al. 2001; Lopez-Schier and St Johnston 2001; Palmer et al. 2014). In other developmental contexts, Notch regulates the survival of cells (Giraldez and Cohen 2003). For example, loss of Notch function results in increased cell death of neuron cells in the mouse nervous system (Mason et al. 2006). Notch signaling has also been associated with cell survival in B-cell malignancies, prostate cancer cells, and myeloma cells (Zhang et al. 2018; Nefedova et al. 2008; Zweidler-McKay et al. 2005).

Notch signaling must be under extremely tight control to keep its proper activity. Emerging evidence indicates that the regulation of Notch signaling seems to be considerately complicated; multiple levels of regulation are added to the pathway via receptor-ligand internalization, posttranslational modification, protein stability, and ligand availability (Kovall et al. 2017). Productive Notch ligand-receptor binding depends on the proper posttranslational modification, such as glycosylation of the receptor (Haines and Irvine 2003). The retention time of Notch and ligands on plasm membrane is determined by the endocytosis of the receptor and ligands (Kandachar and Roegiers 2012), mediated mainly by lysosomal degradation. Mutants that stabilize NICD can cause T-cell acute lymphoblastic leukemia in humans (Grabher et al. 2006). Polarity proteins, such as Numb (Song and Lu 2012) and Crumbs (Nemetschke and Knust 2016), are also required for local distribution of Notch in the plasm membrane, which results in region-specific Notch activity. Like Notch, Dl and Ser are also subject to transmembrane domain cleavage by the γ-secretase complex, with this process called ligand processing, which may be used to downregulate the activity of Notch pathway. Alternatively, ligand processing also could generate biologically soluble ligands that may act as antagonists of Notch signaling (Masuya et al. 2002). Although the mechanisms of signal transduction from the cell surface to the nucleus are relatively simple and clear, it is not fully understood how such a straightforward pathway can result in tremendously complex outcomes in different cellular contexts. Recent studies have revealed that epigenetic modifiers play important roles in regulating Notch activity and may provide a novel angle to explain how and why the various developmental outputs occur in different contexts by a single Notch signaling pathway.

The Advantages of *Drosophila* as Model Organism for Notch Signaling Study

Drosophila melanogaster is an ideal model organism and has been extensively used in scientific research for over 100 years since Professor Thomas H. Morgan (1866–1945), who won the Nobel Prize in Physiology or Medicine in 1933, started to use the *Drosophila* for genetic studies. Owing to several practical advantages that are suitable for laboratory study, *Drosophila* has significantly pushed forward the development of biological research in various fields, such as developmental biology, immunobiology, and metabolism (Mirth et al. 2019). First, *Drosophila* has a short life cycle (about 10 days in the laboratory conditions) and has high fecundity, which allow producing large number of progenies in a short time. Besides, it is relatively easy and cost-effective to maintain the stable *Drosophila* stocks. Second, *Drosophila* has a low number of chromosomes, which make it as one of the most studied organisms in biological research, particularly in genetics and developmental biology (Perrimon 2014; Tolwinski 2017). Notably, the genome sequencing of *Drosophila* reveals that approximately 75% of known human disease-associated genes have counterparts in *Drosophila* (McGurk et al. 2015; Chen and Crowther 2012). The high similarity and conservation in genomic features between *Drosophila* and human enables fly to benefit the biomedical studies of human diseases. Third, a large number of genetic tools are available for *Drosophila* researchers mostly through stock centers, such as the Bloomington Drosophila Stock Center (BDSC) and the Vienna Drosophila Resource Center (VDRC) (Zacharioudaki and Bray 2014; Housden et al. 2014). Large-scale mutagenesis and screen projects are easy to carry out to discover the novel components or novel regulators of a classic pathway. In particular, for those genes that are humongously lethal or semilethal when mutated, somatic or germline clonal analysis based on FRT recombination system would be a good choice. Other FRT system-derived methods, such as MARCM (mosaic analysis with a repressible cell marker) (Lee and Luo 2001), are further developed to create mutant cells in a wild-type background tissue, facilitating reduction of the inter-organismal variability when analyzing mutant versus wild-type tissues.

Notch and its ligands are broadly expressed in many tissues/organs in the *Drosophila*. Therefore, it is of great importance to develop tools to directly examine where the pathway is activated or inhibited. The current arsenal of genetic and biological tools makes *Drosophila* such a valuable model to study the fundamental principles of how Notch signaling transduces the signal and how it is regulated in different cellular contexts, which can deepen the understanding of its roles in physiological and pathological conditions in humans. Table 4.1 shows the commonly used biochemical (antibodies) and genetic (transgenic flies) tools for studying Notch signaling in vitro and in vivo.

Further genetic methods include (1) conditional gene expression and silencing with the Gal4-UAS system, (2) genome-scale bioinformatics analysis, (3) genomic tagging and disruption of genes using CRISPR/Cas9 genome editing for gene

Table 4.1 Commonly used tools for studying Notch signaling in *Drosophila*

Target gene reporters	
E(spl)mβ 1.5-lacZ	Enhancer trap of P-LacZ element in *E(spl)mβ*-locus
vg[BE]-lacZ	Enhancer trap of P-LacZ element in *vg[BE]-lacZ* locus
wg-lacZ	Enhancer trap of P-LacZ element in the *wg* locus
Cut lacZ	Enhancer trap of P-LacZ element in the *cut* locus
Antibodies	
Anti-Notch ICD	Recognizes amino acids 1791–2504 of Notch intracellular domain (C17.9C6, DSHB)
Anti-Delta	Recognizes amino acids 190–833 of Dl protein (C594.9B, DSHB)
Anti-Cut	Recognizes amino acids 1616–1836 of Cut protein (2B10, DSHB)
anti-Wg	Recognizes amino acids 3–468 of Wg protein (4D4, DSHB)
Anti-Hnt	Recognizes amino acids 824–1125 of Peb/Hnt (IG9,DSHB)
Tools to perturb or activate Notch pathway	
N^1	Loss of function of Notch
UAS-NICD	Express Notch intracellular domain under UAS control
UAS-Notch RNAi	RNAi targeting the Notch receptor
H^1	Loss of function of Hairless
$Su(H)^{del47}$	Loss of function of Su(H)
UAS-Dl	Expressing full length Dl under UAS control

All transgenic flies are available from Bloomington *Drosophila* Stock Center. DSHB indicates antibodies available from Developmental Studies Hybridoma Bank, University of Iowa

activation and inactivation (Yu et al. 2013a, 2014), and (4) advanced imaging technology, such as light-sheet microscopy (Lu et al. 2019). All these methods and tools are likely to further facilitate the use of this sophisticated model to better understand the Notch signaling.

It is worth mentioning that Guangzhou Drosophila Stock Center (GDSC), a newly established stock center for generating mutants through genome-wide gene targeting using CRISPR/Cas9 system, has generated more than 1000 mutant stocks. This resource would benefit a lot to those who use *Drosophila* as model for studying Notch signaling and other biological fields.

The Chromatin Assembly Factor CAF-1

Drosophila CAF-1 was first biochemically identified as a chromatin assembly factor about 30 years ago (Smith and Stillman 1989). *Drosophila* genome has three CAF-1-coding genes encoding three subunits, CAF-1 p180, CAF-1 p105, and CAF-1 p55, which correspond to human p150, p60, and p48, respectively (Ridgway and Almouzni 2000) (Table 4.2). Notably, there are two distinct CAF-1 complexes in *Drosophila*, each with three subunits of p180, p105, and p55 or p180, p75, and p55. Among them, p75 is a C-terminally truncated form of p105 in vivo.

Table 4.2 Evolutionarily conserved subunits of CAF-1 complex

Species	Large subunit	Medium subunit	Small subunit
Homo sapiens	p150	p60	p48
Mus musculus	p150	p60	p48
Drosophila melanogaster	p180	pl05	p55
Schmidtea mediterranea	p150	p60	p48
Caenorhabditis elegans	Chaf1	Chaf2	Rba1
Saccharomyces cerevisiae	Cac1	Cac2	Cac3
Arabidopsis thaliana	FAS1	FAS2	MSII

Though p105 and p75 have similar functions, p105 is predominantly expressed during embryogenesis, while p75 dominates after larval formation (Tyler et al. 2001).

CAF-1 has been biochemically well-characterized to be responsible for nucleosome assembly by guiding the histone trafficking and depositing them into chromatin by mediating H3 and H4 dimers onto newly synthesized DNA during DNA replication and DNA repair (Burgess and Zhang 2013). Reduction of CAF-1 activity in culture cells leads to reduced and delayed packaging of the DNA into chromatin, accompanied with DNA replication defects, S-phase arrest, checkpoint activation defects in cell cycle, and even cell death (Klapholz et al. 2009; Jiao et al. 2012; Krude 1995; Collins and Moon 2013). CAF-1 mutant mice exhibit developmental arrest at the embryonic stage with severe alterations in the nuclear organization of constitutive heterochromatin (Houlard et al. 2006). In *Drosophila*, knocking out any of the CAF-1 three subunits results in a similar lethal larval phenotype, and tissue-specific knockdown of CAF-1 p180 in the eye results in eye developmental defects, indicating the CAF-1 complex is also indispensable for the normal development in multicellular organism, including *Drosophila* (Jiao et al. 2012; Song et al. 2007; Huang et al. 2010; Yu et al. 2013b; Wen et al. 2012; Anderson et al. 2011).

However, genes encoding all three subunits CAF-1 are not essential in yeast (Kaufman et al. 1997; Monson et al. 1997; Enomoto and Berman 1998) and plants (Exner et al. 2006; Kirik et al. 2006; Endo et al. 2006; Schonrock et al. 2006), with the CAF-1 mutant in these species being viable, although mutants also exhibit some growth defects. The nonessential role of CAF-1 in unicellular eukaryotes (yeast) and plants appears to be inconsistent with the well-established role of CAF-1 in nucleosome assembly during DNA replication and DNA repair, an activity that might have been expected to be essential for all eukaryotic cells.

Interestingly, emerging evidence reported that CAF-1 p55, the small subunit of *Drosophila* CAF-1, not only functions in the CAF-1 complex but also is a component in several chromatin-modulating complexes, such as PRC1 (Jones et al. 1998) and NuRD (Campbell et al. 2018), indicating that CAF-1 may have multiple functional roles that are not restricted to acting as a histone chaperone. Moreover, it is reasonable to propose that CAF-1 may serve as a protein platform for chromatin metabolism that integrates epigenetic regulation cues of gene transcription by interacting with chromatin modification machinery or transcriptional factors (Yu et al. 2015).

Epigenetic Regulation of Notch Signaling by CAF-1

In a search for which signaling pathway is regulated by CAF-1 in *Drosophila* development, Yu et al. found that tissue-specific knockdown of the *Drosophila* CAF-1 p105, the medium subunit of CAF-1 complex, results in a notched wing phenotype, resembling that of Notch loss-of-function mutations (Yu et al. 2013b). Moreover, the notched wing phenotype could be enhanced by combination with loss of function of Notch, revealing a synergistic genetic interaction between CAF-1 p105 Notch signaling. This study also establishes a functional connection between CAF-1 complex and Notch signaling for the first time.

Similar to wing development, eye development defects caused by eye-specific knockdown of CAF-1 p105 are also significantly enhanced in heterozygous mutant backgrounds of several Notch-positive regulatory components such as Mam (Yu et al. 2013b). Altogether, these results suggest that CAF-1 p105 synergistically interacts with the Notch signaling pathway to regulate normal tissue development.

To further confirm that CAF-1 p105 is required for the normal activity of Notch pathway, Yu et.al generated a null allele of dCAF-1 p105, namely, *CAF-1 p105*^{36}, and performed clonal analyses in wing discs to investigate the effect of dCAF-1 p105 null mutation on Notch signaling, by examination of the developmental defects and the expression change of Notch target genes in the absence of CAF-1 p105. As expected, flies carrying *CAF-1 p105*^{36} clones exhibited a notched wing phenotype, and the protein expression level of *cut* and *wg*, two well-characterized Notch target genes, was also significantly decreased in *CAF-1 p105*^{36} clones, compared with control (Fig. 4.3). The transcriptional level of *cut* is also significantly downregulated when CAF-1 p105 is depleted. These results indicate that the activity of Notch signaling is compromised in the absence of CAF-1 p105 and thus CAF-1 p105 functions as a positive regulator of the Notch signaling pathway by promoting its target gene transcription.

As CAF-1 functions as a histone chaperone, it is likely that its reduction may trigger dilution of newly assembled nucleosomes at key enhancer elements and loosening of chromatin structure, resulting in a more accessible chromatin structure for efficient transcription factor binding to their target loci and activation of key target genes. However, the CAF-1 p105 specifically regulates the output of Notch signaling in the wing disc, since the Hedgehog (Hh) signaling is not affected in CAF-1 p105 mutant clone.

Further, it is found that CAF-1 forms a functional complex with NICD and Su(H), the core transcriptional factor for Notch target gene expression. And this complex directly binds to the enhancer region of one of the Notch target genes, *E(spl)mβ*. The occupancy of Su(H) at Notch target genes is highly increased to efficiently initiate gene transcription when Notch signaling is activated. In the absence of CAF-1 p105, the abundance of Su(H) at the *E(spl)mβ* enhancer region is dramatically decreased, proposing that CAF-1 probably regulates Notch target gene expression, at least in part, by controlling the accessibility and binding abundance of Su(H) to their enhancer regions.

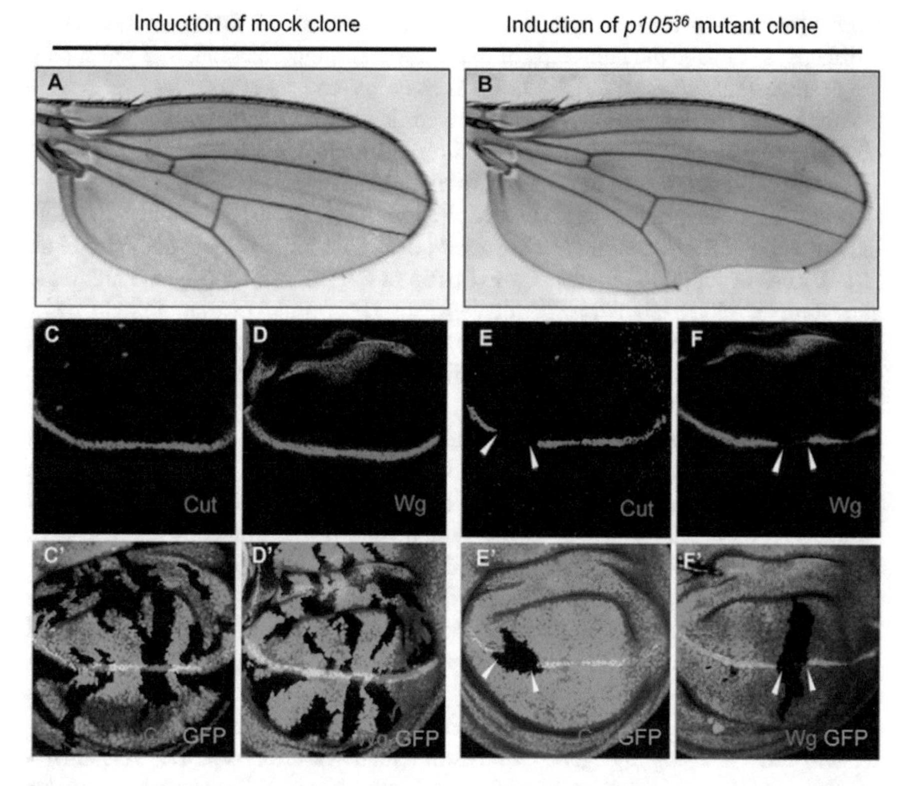

Fig. 4.3 CAF-1-p105 is required for the normal wing development and proper expression of Notch target genes cut and wg. (**a**, **b**) Induction of *CAF-1* $p105^{36}$ mutant clones leads to a notched wing (**b**), whereas induction of mock clones leads to wild-type wings (**a**). (**c**–**f′**) In *CAF-1* $p105^{36}$ mutant clones, the expression of Cut (**e**, **e′**, GFP-negative area, arrowheads) and Wg (**f**, **f′**, FP-negative area, arrowheads) is abolished in a cell-autonomous manner, whereas in the mock clones the expression of both Cut (**c**, **c′**) and Wg (**d**, **d′**) is unaffected. (Images are adapted from Yu et al. (2013b))

It is reported that CAF-1 could act as a chromatin platform that is permissive for transcription by regulating histone modifications by forming complex with other chromatin remodeling complexes (Yu et al. 2015), and histone H4 acetylation is believed to be associated with active promoters of Notch target genes (Giaimo et al. 2018). As expected, H4ac level in the *E(spl)mβ* enhancer region is significantly reduced in CAF-1 p105 mutant flies. These results reveal that CAF-1 p105 promotes Notch target gene expression by maintaining a high level of histone H4 acetylation in the enhancer region of the Notch target genes to establish a local active chromatin structure. Interestingly, CAF-1 function in regulating Notch signaling is dependent on its integrity as a triple subunit complex. Knockdown of any component of CAF-1 complex causes the reduction of *cut* expression and the notched wing phenotype in *Drosophila* (Yu et al. 2013b). However, there are still open questions for how CAF-1 directs the H4 acetylation modification. One possibility is that

CAF-1 recruits histone acetylation machinery, such as p300/Nejire, directly or indirectly, to change the landscape of local chromatin modification, thus enhancing target gene transcription. CAF-1 has functions beyond its classic role in histone assembly and the newly established positive role in Notch signaling in wing development and plays an essential role in proliferating cells.

However, in contrast to the positive role of CAF-1 on Notch target gene expression in wing development, a recent study reported that CAF-1 play a negative role in regulating Notch signaling in *Drosophila* ovarian mitotic follicle cells (Lo et al. 2019). Loss of function of either CAF-1 p105 or CAF-1 p180 caused the increased activation of Notch signaling target genes in *Drosophila* ovarian follicle cells. Further, Notch is functionally responsible for these phenotypes observed in both the CAF-1 p105- and CAF-1 p180-deficient follicle cells. It is still unclear how CAF-1 p180 and CAF-1 p150 suppress Notch target gene expression in mitotic follicle cells. It is likely that CAF-1 have physical interaction with Su(H), which is known to be involved in maintaining the repressive chromatin status for inhibiting Notch target gene expression when it is associated with other repressive subunits, such as Hairless, Gro, or CtBP (Yu et al. 2015; Cheloufi and Hochedlinger 2017; Yuan et al. 2016). Thus, the molecular basis for that CAF-1 play a dual role to sustain cell proliferation positively (in imaginal discs) or negatively (in ovarian follicle cells) may lie in that CAF-1 recruits different histone modification machineries in imaginal discs and follicle cells to regulate *Drosophila* Notch signaling in a tissue context-dependent manner.

Two recent studies in mammals confirmed the negative role of CAF-1 in retrotransposon jumping and gene expression. Hatanaka et al. reported that CAF-1 mediates repressive histone modifications to protect preimplantation mouse embryos from endogenous retrotransposons (Hatanaka et al. 2015). Multiple classes of retrotransposons are derepressed in morula embryos when CAF-1 is depleted, likely through affecting the histone methylation status, thus influencing local chromatin accessibility. The other study found that the p150 and p60, two subunits of mammalian CAF-1 complex, are the most prominent chromatin-modulating factors during transcription factor-mediated reprogramming of mouse fibroblasts to induced pluripotent stem cells (iPS cells) (Cheloufi et al. 2015). Suppression of CAF-1 leads to a more accessible chromatin structure at enhancer elements and the increased binding of Sox2 to pluripotency-specific targets and activation of associated genes during reprogramming (Cheloufi et al. 2015).

Altogether, CAF-1 functions not only as histone chaperone for nucleosome assembly but also as an epigenetic regulation switch for regulating Notch signaling target gene expression in response to integrated proliferation and differentiation signals during animal development. However, CAF-1 does not harbor the histone modification enzyme activity; thus it is likely that CAF-1 works together with the histone modification machinery (histone methylation, histone acetylation, etc.) to regulate Notch signaling activity at the chromatin level through modifying the chromatin structure.

Epigenetic Regulation of Notch Signaling by Other Epigenetic Regulators

In addition to the epigenetic regulation of Notch signaling by the CAF-1 complex, this paragraph briefly summarizes the epigenetic regulation of Notch signaling by other epigenetic regulators. Several epigenetic regulators that are involved in Notch signaling are listed in Table 4.3. Among them, histone acetylation and methylation are main executors for epigenetic regulation of gene transcription (Tchasovnikarova and Kingston 2018).

Table 4.3 Several epigenetic regulators that are involved in Notch signaling

Epigenetic regulators	Molecular activity	Functions	References
UTX	H3K27me3 demethylase	Negatively regulate Notch signaling	Herz et al. (2010)
SIRT1	H4K16 deacetylase	Negatively regulates Notch signaling	Mulligan et al. (2011)
LSD1/KDM1A	H3K4 demethylase	Negatively regulates Notch signaling	Mulligan et al. (2011); Lopez et al. (2016)
CoREST	CoREST complex	Negatively regulates Notch signaling	Lopez et al. (2016)
CoREST	CoREST complex	Positively regulates Notch signaling	Domanitskaya and Schupbach (2012)
HDAC1	Histone deacetylase	Positively regulates Notch signaling	Wang et al. (2018); Mao et al. 2017)
HDAC1	Histone deacetylase	Negatively regulates Notch signaling	Kao et al. (1998); Cunliffe (2004); Yamaguchi et al. (2005); Wu et al. (2016)
Kdm5A	H3K4 demethylase	Negatively regulates Notch signaling	Liefke et al. (2010); Dreval et al. (2019)
Brms1	Histone deacetylase	Positively regulates Notch signaling	Zhang et al. (2014)
Tet2/3	Methylcytosine dioxygenases	Positively regulates Notch signaling	Li et al. (2015)
Nipped-A	SAGA and Tip60	Positively regulates Notch signaling	Gause et al. (2006)
BAP55	SWI/SNF Complex	Positively regulates Notch signaling	Pillidge and Bray (2019)
p300	Histone acetyltransferase	Positively regulates Notch signaling	Franz Oswald et al. (2001)
Tip60	Histone acetyltransferase	Positively regulates Notch signaling	Medgett and Langer (1984)
Nurf 301	NURF complex	Positively regulates Notch signaling	Kugler and Nagel (2010)
Pc	PRC1 complex	Positively regulates Notch signaling	Saj et al. (2010); Tolhuis et al. (2006)

Histone deacetylases are generally associated with transcriptional repressor complexes, such as Sin3 (Barnes et al. 2018), NuRD (Feng and Zhang 2003; Ahringer 2000), and CoREST (You et al. 2001; Domanitskaya and Schupbach 2012) complexes, and have regulatory functions in various signaling pathways. It is generally accepted that HDAC1 forms a transcriptional corepressor complex to modify chromatin structure for target gene silencing. For example, HDAC1 physically interacts with CBF1 (homolog of Su(H) in *Drosophila*), and treatment of HDAC1 inhibitor derepresses Notch target gene *ESR-1* expression in mammalian cells (Kao et al. 1998). In zebrafish, *her4* and *her6*, two of Notch target genes, are upregulated in HDAC1 mutant fish (Cunliffe 2004; Yamaguchi et al. 2005). Furthermore, overexpression of HDAC1 represses the expression of Notch target gene *Hey2* in mice (Wu et al. 2016). In these contexts, HDAC1 negatively regulates Notch signaling. Unexpectedly, opposite to the inhibitory role of HDAC1 in Notch signaling, knockdown of HDAC1 causes a notched wing phenotype and reduces Notch target gene expression in *Drosophila* (Wang et al. 2018), suggesting a positive role of HDAC1 in regulating the Notch pathway during *Drosophila* wing development, although the molecular mechanism behind this remains largely unknown. It is highly possible that HDAC1 directly regulates histone deacetylation status at the Notch target gene locus. Notably, a recent study reported that HDAC1 could activate the Notch signaling pathway to promote metastasis in a similar way (Mao et al. 2017).

The complicated regulation network by other epigenetic regulators, such as LSD1 (Mulligan et al. 2011; Lopez et al. 2016), Brms1 (Zhang et al. 2014), histone acetylase p300 (Franz Oswald et al. 2001), Brahma SWI/SNF chromatin remodeling complex (Pillidge and Bray 2019), UTX (Herz et al. 2010), H3K4 demethylase Kdm5A (Liefke et al. 2010; Dreval et al. 2019), methylcytosine dioxygenases Tet2/3 (Li et al. 2015), SAGA and Tip60 complex (Gause et al. 2006; Medgett and Langer 1984), PcG-TrxG complex (Saj et al. 2010; Tolhuis et al. 2006), Putzig-NURF complex (Kugler and Nagel 2010), and many others, may directly influence Notch-mediated gene transcription activity at the chromatin level and thus explain, at least in part, the pleiotropic effects of Notch in the complex biological processes that affect cell growth, differentiation, and cell death.

Conclusion

The Notch signaling pathway is a highly conserved molecular network that, depending on the cellular context, acts through the regulation of cell proliferation, differentiation, and apoptosis. In order to better control the expression of Notch target gene expression, the Notch signaling must be precisely regulated at different steps in a series of developmental events. Epigenetic regulation of Notch signaling by CAF-1 and other epigenetic regulators plays essential roles in fine-tuning the transcriptional output of Notch signaling to coordinate multicellular organism development. It remains an open question as to why and how different epigenetic regulators are involved in mediating different histone modifications status, leading to different transcriptional outputs of either gene repression or gene activation in one specific signal transduction pathway.

References

Ahringer J (2000) NuRD and SIN3 histone deacetylase complexes in development. Trends Genet 16(8):351–356

Alabi RO, Farber G, Blobel CP (2018) Intriguing roles for endothelial ADAM10/Notch signaling in the development of organ-specific vascular beds. Physiol Rev 98(4):2025–2061

Anderson AE et al (2011) The enhancer of trithorax and polycomb gene Caf1/p55 is essential for cell survival and patterning in Drosophila development. Development 138(10):1957–1966

Barnes VL et al (2018) Systematic analysis of SIN3 histone modifying complex components during development. Sci Rep 8(1):17048

Borggrefe T, Liefke R (2012) Fine-tuning of the intracellular canonical Notch signaling pathway. Cell Cycle 11(2):264–276

Borggrefe T, Oswald F (2016) Setting the stage for Notch: the Drosophila Su(H)-hairless repressor complex. PLoS Biol 14(7):e1002524

Bray SJ (2016) Notch signalling in context. Nat Rev Mol Cell Biol 17(11):722–735

Bugeon L et al (2011) The NOTCH pathway contributes to cell fate decision in myelopoiesis. Haematologica 96(12):1753–1760

Burgess RJ, Zhang Z (2013) Histone chaperones in nucleosome assembly and human disease. Nat Struct Mol Biol 20(1):14–22

Campbell AE et al (2018) NuRD and CAF-1-mediated silencing of the D4Z4 array is modulated by DUX4-induced MBD3L proteins. Elife 7:pii: e31023

Cheloufi S, Hochedlinger K (2017) Emerging roles of the histone chaperone CAF-1 in cellular plasticity. Curr Opin Genet Dev 46:83–94

Cheloufi S et al (2015) The histone chaperone CAF-1 safeguards somatic cell identity. Nature 528(7581):218–224

Chen KF, Crowther DC (2012) Functional genomics in Drosophila models of human disease. Brief Funct Genomics 11(5):405–415

Collins H, Moon NS (2013) The components of Drosophila histone chaperone dCAF-1 are required for the cell death phenotype associated with rbf1 mutation. G3 (Bethesda) 3(10):1639–1647

Cunliffe VT (2004) Histone deacetylase 1 is required to repress Notch target gene expression during zebrafish neurogenesis and to maintain the production of motoneurones in response to hedgehog signalling. Development 131(12):2983–2995

De Strooper B et al (1999) A presenilin-1-dependent gamma-secretase-like protease mediates release of Notch intracellular domain. Nature 398(6727):518–522

Deng WM, Althauser C, Ruohola-Baker H (2001) Notch-Delta signaling induces a transition from mitotic cell cycle to endocycle in Drosophila follicle cells. Development 128(23):4737–4746

Domanitskaya E, Schupbach T (2012) CoREST acts as a positive regulator of Notch signaling in the follicle cells of Drosophila melanogaster. J Cell Sci 125(Pt 2):399–410

Dreval K, Lake RJ, Fan HY (2019) HDAC1 negatively regulates selective mitotic chromatin binding of the Notch effector RBPJ in a KDM5A-dependent manner. Nucleic Acids Res 47(9):4521–4538

Endo M et al (2006) Increased frequency of homologous recombination and T-DNA integration in Arabidopsis CAF-1 mutants. EMBO J 25(23):5579–5590

Enomoto S, Berman J (1998) Chromatin assembly factor I contributes to the maintenance, but not the re-establishment, of silencing at the yeast silent mating loci. Genes Dev 12(2):219–232

Exner V et al (2006) Chromatin assembly factor CAF-1 is required for cellular differentiation during plant development. Development 133(21):4163–4172

Feng Q, Zhang Y (2003) The NuRD complex: linking histone modification to nucleosome remodeling. Curr Top Microbiol Immunol 274:269–290

Franz Oswald BT, Dobner T, Bourteele S, Kostezka U, Adler G, Liptay S, Schmid RM (2001) p300 acts as a transcriptional coactivator for mammalian Notch-1. Mol Cell Biol 21(22):7761–7774

Gause M et al (2006) Nipped-A, the Tra1/TRRAP subunit of the Drosophila SAGA and Tip60 complexes, has multiple roles in Notch signaling during wing development. Mol Cell Biol 26(6):2347–2359

Genethliou N et al (2009) SOX1 links the function of neural patterning and Notch signalling in the ventral spinal cord during the neuron-glial fate switch. Biochem Biophys Res Commun 390(4):1114–1120

Giaimo BD et al (2018) Histone variant H2A.Z deposition and acetylation directs the canonical Notch signaling response. Nucleic Acids Res 46(16):8197–8215

Giraldez AJ, Cohen SM (2003) Wingless and Notch signaling provide cell survival cues and control cell proliferation during wing development. Development 130(26):6533–6543

Grabher C, von Boehmer H, Look AT (2006) Notch 1 activation in the molecular pathogenesis of T-cell acute lymphoblastic leukaemia. Nat Rev Cancer 6(5):347–359

Haines N, Irvine KD (2003) Glycosylation regulates Notch signalling. Nat Rev Mol Cell Biol 4(10):786–797

Hatanaka Y et al (2015) Histone chaperone CAF-1 mediates repressive histone modifications to protect preimplantation mouse embryos from endogenous retrotransposons. Proc Natl Acad Sci U S A 112(47):14641–14646

Herz HM et al (2010) The H3K27me3 demethylase dUTX is a suppressor of Notch- and Rb-dependent tumors in Drosophila. Mol Cell Biol 30(10):2485–2497

Houlard M et al (2006) CAF-1 is essential for heterochromatin organization in pluripotent embryonic cells. PLoS Genet 2(11):e181

Housden BE, Li J, Bray SJ (2014) Visualizing Notch signaling in vivo in Drosophila tissues. Methods Mol Biol 1187:101–113

Huang H et al (2010) Drosophila CAF-1 regulates HP1-mediated epigenetic silencing and pericentric heterochromatin stability. J Cell Sci 123(Pt 16):2853–2861

Jiao R-J, Wu Q-H, Liu J-Y, Chen Y-X (2012) dCAF-1-p55 is essential for Drosophila development and involved in the maintenance of chromosomal stability. Prog Biochem Biophys 39(11):1073–1081

Jones CA et al (1998) The Drosophila esc and E(z) proteins are direct partners in polycomb group-mediated repression. Mol Cell Biol 18(5):2825–2834

Kandachar V, Roegiers F (2012) Endocytosis and control of Notch signaling. Curr Opin Cell Biol 24(4):534–540

Kao HY et al (1998) A histone deacetylase corepressor complex regulates the Notch signal transduction pathway. Genes Dev 12(15):2269–2277

Kaufman PD, Kobayashi R, Stillman B (1997) Ultraviolet radiation sensitivity and reduction of telomeric silencing in Saccharomyces cerevisiae cells lacking chromatin assembly factor-I. Genes Dev 11(3):345–357

Kirik A et al (2006) The chromatin assembly factor subunit FASCIATA1 is involved in homologous recombination in plants. Plant Cell 18(10):2431–2442

Klapholz B et al (2009) CAF-1 is required for efficient replication of euchromatic DNA in Drosophila larval endocycling cells. Chromosoma 118(2):235–248

Klueg KM, Muskavitch MA (1999) Ligand-receptor interactions and trans-endocytosis of Delta, Serrate and Notch: members of the Notch signalling pathway in Drosophila. J Cell Sci 112(Pt 19):3289–3297

Kovall RA et al (2017) The canonical Notch signaling pathway: structural and biochemical insights into shape, sugar, and force. Dev Cell 41(3):228–241

Krude T (1995) Chromatin assembly factor 1 (CAF-1) colocalizes with replication foci in HeLa cell nuclei. Exp Cell Res 220(2):304–311

Kugler SJ, Nagel AC (2010) A novel Pzg-NURF complex regulates Notch target gene activity. Mol Biol Cell 21(19):3443–3448

Lee T, Luo L (2001) Mosaic analysis with a repressible cell marker (MARCM) for Drosophila neural development. Trends Neurosci 24(5):251–254

Li C et al (2015) Overlapping requirements for Tet2 and Tet3 in Normal development and hematopoietic stem cell emergence. Cell Rep 12(7):1133–1143

Liefke R et al (2010) Histone demethylase KDM5A is an integral part of the core Notch-RBP-J repressor complex. Genes Dev 24(6):590–601

Lo PK et al (2019) Inhibition of Notch signaling by the p105 and p180 subunits of Drosophila chromatin assembly factor 1 is required for follicle cell proliferation. J Cell Sci 132(2):jcs224170

Lopez CI et al (2016) The chromatin modifying complex CoREST/LSD1 negatively regulates notch pathway during cerebral cortex development. Dev Neurobiol 76(12):1360–1373

Lopez-Schier H, St Johnston D (2001) Delta signaling from the germ line controls the proliferation and differentiation of the somatic follicle cells during Drosophila oogenesis. Genes Dev 15(11):1393–1405

Lu CH et al (2019) Lightsheet localization microscopy enables fast, large-scale, and three-dimensional super-resolution imaging. Commun Biol 2:177

Mao G, Jin H, Wu L (2017) DDX23-Linc00630-HDAC1 axis activates the Notch pathway to promote metastasis. Oncotarget 8(24):38937–38949

Mason HA et al (2006) Loss of notch activity in the developing central nervous system leads to increased cell death. Dev Neurosci 28(1–2):49–57

Masuya M et al (2002) The soluble Notch ligand, Jagged-1, inhibits proliferation of CD34+ macrophage progenitors. Int J Hematol 75(3):269–276

McGurk L, Berson A, Bonini NM (2015) Drosophila as an in vivo model for human neurodegenerative disease. Genetics 201(2):377–402

Medgett IC, Langer SZ (1984) Heterogeneity of smooth muscle alpha adrenoceptors in rat tail artery in vitro. J Pharmacol Exp Ther 229(3):823–830

Mirth CK, Nogueira Alves A, Piper MD (2019) Turning food into eggs: insights from nutritional biology and developmental physiology of Drosophila. Curr Opin Insect Sci 31:49–57

Monson EK, de Bruin D, Zakian VA (1997) The yeast Cac1 protein is required for the stable inheritance of transcriptionally repressed chromatin at telomeres. Proc Natl Acad Sci U S A 94(24):13081–13086

Mulligan P et al (2011) A SIRT1-LSD1 corepressor complex regulates Notch target gene expression and development. Mol Cell 42(5):689–699

Muskavitch MA (1994) Delta-Notch signaling and Drosophila cell fate choice. Dev Biol 166(2):415–430

Mutvei AP, Fredlund E, Lendahl U (2015) Frequency and distribution of Notch mutations in tumor cell lines. BMC Cancer 15:311

Nefedova Y et al (2008) Inhibition of Notch signaling induces apoptosis of myeloma cells and enhances sensitivity to chemotherapy. Blood 111(4):2220–2229

Nemetschke L, Knust E (2016) Drosophila crumbs prevents ectopic Notch activation in developing wings by inhibiting ligand-independent endocytosis. Development 143(23):4543–4553

Ntziachristos P et al (2014) From fly wings to targeted cancer therapies: a centennial for Notch signaling. Cancer Cell 25(3):318–334

Palmer WH, Jia D, Deng WM (2014) Cis-interactions between Notch and its ligands block ligand-independent Notch activity. Elife 3. https://doi.org/10.7554/eLife.04415

Perrimon N (2014) Drosophila developmental biology methods. Methods 68(1):1

Pillidge Z, Bray SJ (2019) SWI/SNF chromatin remodeling controls Notch-responsive enhancer accessibility. EMBO Rep 20(5):pii: e46944

Radtke F, Wilson A, MacDonald HR (2005) Notch signaling in hematopoiesis and lymphopoiesis: lessons from Drosophila. BioEssays 27(11):1117–1128

Ridgway P, Almouzni G (2000) CAF-1 and the inheritance of chromatin states: at the crossroads of DNA replication and repair. J Cell Sci 113(Pt 15):2647–2658

Saj A et al (2010) A combined ex vivo and in vivo RNAi screen for notch regulators in Drosophila reveals an extensive notch interaction network. Dev Cell 18(5):862–876

Salazar JL, Yamamoto S (2018) Integration of Drosophila and human genetics to understand notch signaling related diseases. Adv Exp Med Biol 1066:141–185

Schonrock N et al (2006) Functional genomic analysis of CAF-1 mutants in Arabidopsis thaliana. J Biol Chem 281(14):9560–9568

Shepard SB, Broverman SA, Muskavitch MA (1989) A tripartite interaction among alleles of Notch, Delta, and Enhancer of split during imaginal development of Drosophila melanogaster. Genetics 122(2):429–438

Siren M, Portin P (1989) Interaction of hairless, delta, enhancer of split and notch genes of Drosophila melanogaster as expressed in adult morphology. Genet Res 54(1):23–26

Smith S, Stillman B (1989) Purification and characterization of CAF-I, a human cell factor required for chromatin assembly during DNA replication in vitro. Cell 58(1):15–25

Song Y, Lu B (2012) Interaction of Notch signaling modulator Numb with alpha-Adaptin regulates endocytosis of Notch pathway components and cell fate determination of neural stem cells. J Biol Chem 287(21):17716–17728

Song Y et al (2007) CAF-1 is essential for Drosophila development and involved in the maintenance of epigenetic memory. Dev Biol 311(1):213–222

Tchasovnikarova IA, Kingston RE (2018) Beyond the histone code: a physical map of chromatin states. Mol Cell 69(1):5–7

Thomas U, Speicher SA, Knust E (1991) The Drosophila gene Serrate encodes an EGF-like transmembrane protein with a complex expression pattern in embryos and wing discs. Development 111(3):749–761

Tolhuis B et al (2006) Genome-wide profiling of PRC1 and PRC2 Polycomb chromatin binding in Drosophila melanogaster. Nat Genet 38(6):694–699

Tolwinski NS (2017) Introduction: Drosophila-A model system for developmental biology. J Dev Biol 5(3):9

Tyler JK et al (2001) Interaction between the Drosophila CAF-1 and ASF1 chromatin assembly factors. Mol Cell Biol 21(19):6574–6584

Wang Z et al (2018) The histone deacetylase HDAC1 positively regulates Notch signaling during Drosophila wing development. Biol Open 7(2):bio029637

Welshons WJ (1958a) A preliminary investigation of pseudoallelism at the notch locus of Drosophila melanogaster. Proc Natl Acad Sci U S A 44(3):254–258

Welshons WJ (1958b) The analysis of a pseudoallelic recessive lethal system at the notch locus of Drosophila melanogaster. Cold Spring Harb Symp Quant Biol 23:171–176

Wen P, Quan Z, Xi R (2012) The biological function of the WD40 repeat-containing protein p55/Caf1 in Drosophila. Dev Dyn 241(3):455–464

Wu LM et al (2016) Zeb2 recruits HDAC-NuRD to inhibit Notch and controls Schwann cell differentiation and remyelination. Nat Neurosci 19(8):1060–1072

Yamaguchi M et al (2005) Histone deacetylase 1 regulates retinal neurogenesis in zebrafish by suppressing Wnt and Notch signaling pathways. Development 132(13):3027–3043

Yang G et al (2019) Structural basis of Notch recognition by human gamma-secretase. Nature 565(7738):192–197

You A et al (2001) CoREST is an integral component of the CoREST- human histone deacetylase complex. Proc Natl Acad Sci U S A 98(4):1454–1458

Yu Z et al (2013a) Highly efficient genome modifications mediated by CRISPR/Cas9 in Drosophila. Genetics 195(1):289–291

Yu Z et al (2013b) CAF-1 promotes Notch signaling through epigenetic control of target gene expression during Drosophila development. Development 140(17):3635–3644

Yu Z et al (2014) Various applications of TALEN- and CRISPR/Cas9-mediated homologous recombination to modify the Drosophila genome. Biol Open 3(4):271–280

Yu Z et al (2015) Histone chaperone CAF-1: essential roles in multi-cellular organism development. Cell Mol Life Sci 72(2):327–337

Yuan Z et al (2016) Structure and function of the Su(H)-hairless repressor complex, the major antagonist of Notch signaling in Drosophila melanogaster. PLoS Biol 14(7):e1002509
Zacharioudaki E, Bray SJ (2014) Tools and methods for studying Notch signaling in Drosophila melanogaster. Methods 68(1):173–182
Zhang Q et al (2014) dBrms1 acts as a positive regulator of Notch signaling in Drosophila wing. J Genet Genomics 41(6):317–325
Zhang R, Engler A, Taylor V (2018) Notch: an interactive player in neurogenesis and disease. Cell Tissue Res 371(1):73–89
Zweidler-McKay PA et al (2005) Notch signaling is a potent inducer of growth arrest and apoptosis in a wide range of B-cell malignancies. Blood 106(12):3898–3906

Chapter 5
Regulation of Proneural Wave Propagation Through a Combination of Notch-Mediated Lateral Inhibition and EGF-Mediated Reaction Diffusion

Makoto Sato and Tetsuo Yasugi

Abstract Notch-mediated lateral inhibition regulates binary cell fate choice, resulting in salt-and-pepper pattern formation during various biological processes. In many cases, Notch signaling acts together with other signaling systems. However, it is not clear what happens when Notch signaling is combined with other signaling systems. Mathematical modeling and the use of a simple biological model system will be essential to address this uncertainty. A wave of differentiation in the *Drosophila* visual center, the "proneural wave," accompanies the activity of the Notch and EGF signaling pathways. Although all of the Notch signaling components required for lateral inhibition are involved in the proneural wave, no salt-and-pepper pattern is found during the progression of the proneural wave. Instead, Notch is activated along the wave front and regulates proneural wave progression. How does Notch signaling control wave propagation without forming a salt-and-pepper pattern? A mathematical model of the proneural wave, based on biological evidence, has demonstrated that Notch-mediated lateral inhibition is implemented within the proneural wave and that the diffusible action of EGF cancels salt-and-pepper pattern formation. The results from numerical simulation have been confirmed by genetic experiments in vivo and suggest that the combination of Notch-mediated lateral inhibition and EGF-mediated reaction diffusion enables a novel function of Notch signaling that regulates propagation of the proneural wave. Similar mechanisms may play important roles in diverse biological processes found in animal development and cancer pathogenesis.

M. Sato (✉)
Mathematical Neuroscience Unit, Institute for Frontier Science Initiative, Kanazawa University, Kanazawa-shi, Ishikawa, Japan

Laboratory of Developmental Neurobiology, Graduate School of Medical Sciences, Kanazawa University, Kanazawa-shi, Ishikawa, Japan
e-mail: makotos@staff.kanazawa-u.ac.jp

T. Yasugi
Mathematical Neuroscience Unit, Institute for Frontier Science Initiative, Kanazawa University, Kanazawa-shi, Ishikawa, Japan

J. Reichrath, S. Reichrath (eds.), *Notch Signaling in Embryology and Cancer*, Advances in Experimental Medicine and Biology 1218,
https://doi.org/10.1007/978-3-030-34436-8_5

Keywords Notch · Delta · Lateral inhibition · EGF · Reaction diffusion · JAK/STAT · Noise resistance · Drosophila · Visual system · Proneural wave · Mathematical model

The Proneural Wave

Notch is evolutionarily conserved throughout the animal kingdom and regulates various biological processes (Artavanis-Tsakonas et al. 1999). In many cases, the Notch receptor is activated by binding to its membrane-bound ligand, Delta, and represses the expression of target genes, including Delta itself (Fig. 5.1a, b). Since Delta activates Notch on neighboring cells, these molecules form a feedback loop between adjacent cells and establish a binary cell fate through a process called lateral inhibition (Simpson 1990; Collier et al. 1996). Notch-mediated lateral inhibition plays an essential role in specifying differentiated cells from a group of undifferentiated cells in a spatially regulated manner by forming a salt-and-pepper pattern (Fig. 5.1c).

Notch-mediated lateral inhibition was discovered through a series of studies focusing on sensory bristles, which are part of the peripheral nervous system of *Drosophila melanogaster* (Simpson 1990). Individual sensory bristles are formed from sensory organ precursors (SOPs), which are differentiated from undifferentiated epithelial cells by the function of proneural transcription factors of the Achaete-Scute complex (AS-C) (Fig. 5.1d) (Ghysen et al. 1993). Although AS-C expression is widely upregulated in sheets of epithelial cells, only a small number of cells become SOPs as a result of Notch-mediated lateral inhibition. In differentiating epithelial cells, AS-C upregulates Delta expression, which then represses AS-C expression in adjacent cells through Notch signaling (Fig. 5.1a, b) (Kunisch et al. 1994). AS-C expression is downregulated in cells in which Notch signaling is activated and Delta expression is reduced. Furthermore, Delta autonomously represses Notch function in differentiating cells through *cis*-inhibition (Fig. 5.1a, b) (del Alamo et al. 2011; Sprinzak et al. 2010). Thus, Delta, Notch, and AS-C form a feedback loop between adjacent cells that enables a binary cell fate decision. As a result, a small number of epithelial cells are selected as SOPs, while the surrounding cells are kept undifferentiated, forming a salt-and-pepper pattern (Fig. 5.1c).

Notch-mediated lateral inhibition is also widely conserved in other developmental processes. However, Notch signaling often cooperates with other signaling systems and shows behavior that is complex in comparison with that in the classic case of sensory bristle formation (Doroquez and Rebay 2006; Dutt et al. 2004; Sundaram 2005). Within the diverse repertoire of such developmental processes, the waves of differentiation found in the developing eye and brain of *Drosophila* are unique examples (Heberlein et al. 1993; Ma et al. 1993; Sato et al. 2013; Yasugi et al. 2008).

The development of the retina in flies, fish, and chickens includes waves of differentiation that accompany Notch and secreted factors, such as Hedgehog (Hh), that trigger neural differentiation (Heberlein et al. 1993; Ma et al. 1993; Neumann

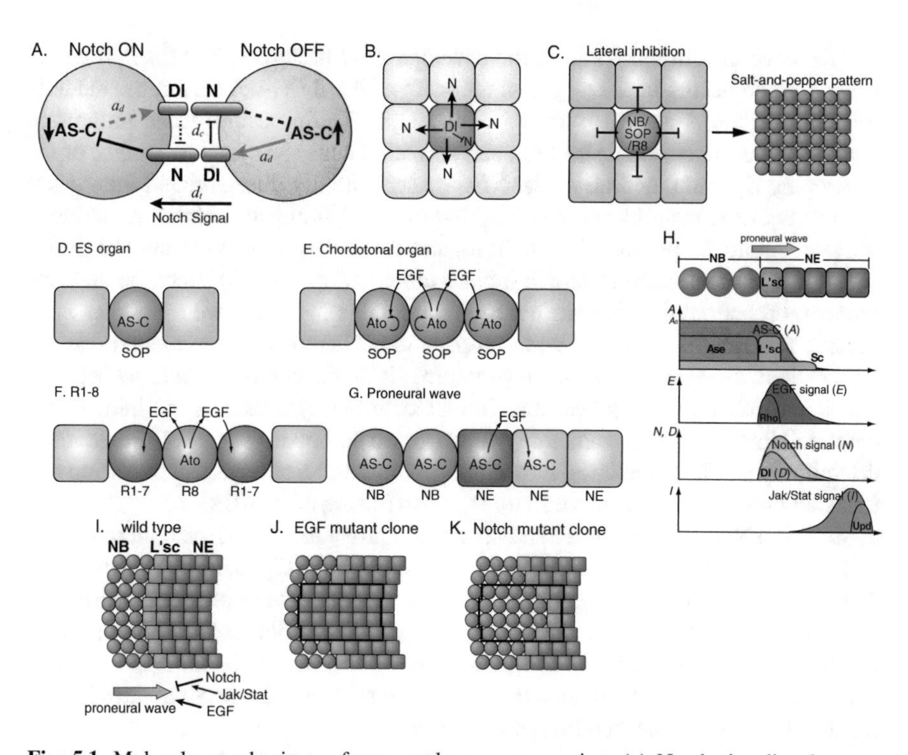

Fig. 5.1 Molecular mechanisms of proneural wave progression. (**a**) Notch signaling between adjacent cells mediates the binary cell fate decision. (**b**) The differentiating cells prevent the neighboring cells from differentiating through Delta/Notch signaling. (**c**) The process of lateral inhibition establishes the formation of salt-and-pepper patterns. (**d–g**) Neural differentiation under the control of proneural transcription factors, AS-C and Ato, and diffusible EGF. Sensory organ precursor (SOP) differentiation of external sensory (ES) organs does not propagate (**d**). Induction of Ato autoregulation by EGF regulates the propagation of chordotonal SOP differentiation (**e**). Since EGF does not induce Ato expression, R1–8 differentiation does not propagate in the retina (**f**). Induction of AS-C expression by EGF regulates proneural wave propagation (**g**). (**h–k**) The proneural wave sweeps across the neuroepithelial (NE) sheet (*blue*). The NE cells express L'sc (*green*), triggering the differentiation from NEs into neuroblasts (NBs) (*red*). (**h**) Schema showing the relative distributions of the AS-C family (L'sc, Sc, and Ase), Rho and Dl expression, and the activities of the EGF and Notch signaling pathways. (**i**) Wild type. (**j, k**) In clones mutant for EGF and Notch signaling pathways, the proneural wave is eliminated (**j**) and accelerated (**k**), respectively

and Nuesslein-Volhard 2000; Yang et al. 2009). Previous mathematical models have revealed essential roles of Notch signaling in the formation of a salt-and-pepper pattern of photoreceptor neurons (Formosa-Jordan et al. 2012; Lubensky et al. 2011; Pennington and Lubensky 2010). The mathematical model of neural differentiation in the retina suggests that Delta restricts the propagation of the differentiation wave (Formosa-Jordan et al. 2012). However, Notch signaling plays an early role in undifferentiated cells to enable them to acquire a proneural state prior to wave progression in the fly retina (Baonza and Freeman 2001; Baker and Yu 1997). Therefore, it is technically difficult to address this hypothesis by using the fly retina.

The wave of differentiation in the developing fly brain is called the "proneural wave," which occurs in the largest component of the fly visual center, the medulla (Fig. 5.1i) (Sato et al. 2013; Yasugi et al. 2008). The proneural wave progresses along the two-dimensional sheet of neuroepithelial cells (NEs) on the surface of the developing fly brain, and all of the NEs are sequentially differentiated into neural stem cells called neuroblasts (NBs) behind the proneural wave. Notch signaling is activated along the proneural wave front, and the wave progression and NB differentiation are accelerated in Notch mutant clones (Fig. 5.1h, k). Because an early proneural function of Notch has not been reported in the brain, we can focus on its specific function in proneural wave propagation. Therefore, the proneural wave is an excellent model system with which to investigate the complex interplay between Notch and other signaling systems. To elucidate the dynamics of the interactions between Notch and other signaling pathways, we and others have performed interdisciplinary studies combining molecular genetics and mathematical modeling and focusing on the proneural wave (Sato et al. 2016; Jorg et al. 2019).

EGF and Notch signaling are activated at the proneural wave front and positively and negatively regulate wave progression, respectively (Fig. 5.1h, j) (Yasugi et al. 2010). EGF signaling is evidently essential for proneural wave progression because wave progression is completely abolished when EGF signaling is blocked (Fig. 5.1j). Because of the evolutionarily conserved genetic cassette containing AS-C, Delta, and Notch, it is natural to assume that the proneural wave also accompanies Notch-mediated lateral inhibition (Ghysen et al. 1993).

However, the roles of Notch signaling in the proneural wave are very unclear because there is no salt-and-pepper pattern during proneural wave progression (Fig. 5.1i) (Reddy et al. 2010; Yasugi et al. 2008, 2010). Notch signaling appears to regulate the speed of proneural wave propagation instead of generating the salt-and-pepper pattern, because wave progression is accelerated in Notch mutant clones (Fig. 5.1k). How does Notch control wave propagation without forming a salt-and-pepper pattern?

In this book chapter, we summarize the recent advances in our understanding of the function of Notch signaling within a complex system that includes multiple signal transduction pathways. Although we mainly focus on the proneural wave in the developing fly brain, we believe that the essential mechanisms are conserved in other biological systems.

The Link Between Proneural Transcription Factors and Notch Signaling

The cell differentiation process of the proneural wave is highly analogous to that of SOP development, in which SOPs, which are neural progenitor cells, are selected from a group of undifferentiated epithelial cells by the expression of proneural transcription factors in the AS-C, including Achaete (Ac), Scute (Sc), Lethal of Scute

(L'sc), and Asense (Ase) (Ghysen et al. 1993). These factors all belong to the basic helix-loop-helix (bHLH) transcription factor family. Delta expression in differentiating SOPs under the control of AS-C represses neuronal differentiation in the surrounding cells. Therefore, only a small number of epithelial cells are selected as SOPs showing a salt-and-pepper pattern (Fig. 5.1c) (Ghysen et al. 1993; Kunisch et al. 1994; Corson et al. 2017).

During proneural wave progression, sheet-like NEs sequentially differentiate into NBs as a result of the action of AS-C family proteins (Fig. 5.1g, h). The transient expression of L'sc among these proteins defines the wave front in which NEs differentiate into NBs. AS-C proteins, including L'sc, act redundantly to trigger differentiation (Yasugi et al. 2008).

Although L'sc expression is strikingly restricted to the wave front cells, the specific function of L'sc is unclear, because small deletions of *l'sc* together with a subset of other AS-C family genes show only minor delay in NB differentiation (Yasugi et al. 2008). Thus, the functions of the AS-C genes are highly redundant. When all AS-C genes are deleted, wave progression is significantly delayed, but NBs are eventually differentiated (Yasugi et al. 2008). All of the AS-C transcription factors need to dimerize with another bHLH transcription factor, Daughterless (Da), to regulate target gene transcription (Cabrera and Alonso 1991). Indeed, wave progression is significantly delayed in *da* mutant cells. However, NBs are eventually differentiated in the absence of *da* (Yasugi et al. 2008). Thus, there are unidentified additional mechanisms that trigger NB differentiation in the absence of AS-C function.

Consistent with the function of Delta and Notch in SOP development, Delta is expressed in L'sc-positive differentiating NEs at the wave front (Fig. 5.1h). Thus, L'sc and other AS-C proteins most likely upregulate Delta expression at the wave front. As a result, Notch signaling is activated in nearby NEs. Thus, the relationship between AS-C, Delta, and Notch is highly conserved between SOP development and proneural wave progression (Fig. 5.1a).

The Link Between Proneural Transcription Factors and EGF Signaling

The most prominent characteristic of the proneural wave is its propagation along the NE sheet, which requires the involvement of EGF signaling (Yasugi et al. 2010). EGF plays a central role in controlling proneural wave progression, because the EGF ligand is produced and secreted from L'sc-positive wave front cells and activates EGF signaling in neighboring NEs. Additionally, EGF signaling triggers NB differentiation through expression of AS-C proteins, including L'sc, and EGF signaling is mandatory for proneural wave propagation (Fig. 5.1g, h). In the absence of the EGF signal, NB differentiation is completely blocked and proneural wave progression is terminated (Fig. 5.1j).

EGF signaling plays essential roles during the development of SOPs in the chordotonal organ, a stretch receptor (Fig. 5.1e). Chordotonal SOPs use Atonal (Ato) as a bHLH proneural transcription factor, which also dimerizes with Da and triggers SOP differentiation (Jarman et al. 1993). Under the control of Ato, SOPs express the EGF ligand, which then triggers differentiation of surrounding cells as secondary SOPs by augmenting the autoregulation of Ato (Fig. 5.1e) (zur Lage et al. 2003). The link between EGF and Ato enables the propagation of chordotonal SOP differentiation in the leg disc, a phenomenon very similar to the proneural wave.

Similar developmental processes occur during photoreceptor differentiation in the fly retina. Ato triggers the differentiation of the primary photoreceptor R8 and recruits the differentiation of the other photoreceptors, R1–7, through EGF signaling (Fig. 5.1f) (Jarman et al. 1994; Freeman 1996). In the retina, the progression of the wave of differentiation is essentially driven by Hh (Heberlein et al. 1993; Ma et al. 1993). Although EGF signaling is involved in the differentiation of R1–7 cells adjacent to R8, Ato function is not cell autonomously required for R1–7 differentiation (Jarman et al. 1994; Freeman 1996). In contrast to the case of chordotonal SOPs, EGF signaling does not trigger R1–7 differentiation through Ato in the retina. As a result, R1–7 differentiation does not propagate and is restricted to the cells near R8. These observations suggest that EGF signaling alone is not sufficient for propagation of the wave of differentiation (Fig. 5.1f).

The proneural wave may involve a mechanism similar to that involved in the propagation of chordotonal SOPs. Instead of Ato, AS-C may be linked with EGF signaling. Since EGF signaling has been shown to trigger NB differentiation by upregulating AS-C proneural transcription factors (Yasugi et al. 2010), we assume that EGF directly activates AS-C expression (Fig. 5.1g). In *Drosophila*, one of the major EGF ligands is Spitz (Spi), a membrane-bound ligand that is cleaved by the metalloprotease Rhomboid (Rho). Upon Rho expression, membrane-bound Spi (mSpi) is processed into the secreted active form, sSpi, which then binds to EGF receptor (EGFR) and activates EGF signaling (Urban et al. 2001). We assume that AS-C activates EGF signaling by upregulating the transcription of *Rho*, because inactivation of AS-C function results in reductions in EGF signaling (Sato et al. 2016). Thus, mutual regulation between AS-C and EGF signaling is the core mechanism of proneural wave propagation (Fig. 5.1g).

Establishing a Mathematical Model of the Proneural Wave

Extensive genetic screening has revealed that the Jak/Stat and Hippo signaling pathways play essential roles in proneural wave progression in addition to the EGF and Notch pathways (Yasugi et al. 2008; Kawamori et al. 2011). We assume that we have already identified all of the signaling systems essential for proneural wave progression. Among these four signaling systems, the EGF and Notch systems are activated at the wave front (Fig. 5.1h). Therefore, we initially focused on the roles of EGF and Notch.

As discussed above, Notch-mediated lateral inhibition established by the feedback loop of AS-C, Delta, and Notch is evolutionarily conserved and is observed in many biological processes (Ghysen et al. 1993). Therefore, this feedback loop should be conserved in the proneural wave.

On the basis of these biological observations, we established a mathematical model containing four variables (Fig. 5.2a, b). In this model, E is a composite variable for the EGF ligand concentration and EGF signaling. Note that the behavior of the model is essentially the same when diffusion of the EGF ligand and activation of EGF signaling are considered separately. The rate of change in E is influenced by its diffusion ($d_e \Delta E$; $\Delta = (\partial^2/\partial x^2) + (\partial^2/\partial y^2)$) because the EGF ligand diffuses and activates EGF signaling in surrounding cells. E is reduced by degradation ($k_e E$) because all molecules are degraded, and EGF ligands and EGF signaling components should also be degraded proportionally to E.

The variables N and D are Notch signaling activity and Delta expression, respectively, and are reduced by degradation ($k_n N$ and $k_d D$). When we consider the lateral inhibition mediated by Notch and Delta, Notch signaling in the ith and jth cell ($N_{i,j}$) is activated by Delta expressed in the adjacent lth and mth cells (*trans*-Dl; $d_t D_{l,m}$) and inhibited by Delta expressed in the same ith and jth cell (*cis*-Dl; $d_c D_{i,j}$) (del Alamo et al. 2011; Sprinzak et al. 2010). When *cis*-inhibition is stronger than *trans*-activation, $N_{i,j}$ may have a negative value. The positive and negative values of Notch signaling correspond to the functions of the transcription factor Su(H) as an activator and repressor downstream of Notch signaling, respectively (Yuan et al. 2016). Alternatively, the term of *cis*-inhibition can be $-d_c D_{i,j} N_{i,j}$ to maintain $N_{i,j}$ nonnegative (Tanaka et al. 2018).

A is a relatively abstract variable indicating the state of differentiation ($A = 0$ in undifferentiated NEs, $A = 1$ in differentiated NBs), which is closely related to the expression levels of AS-C proteins, including L'sc, Sc, and Ase. These proteins are expressed near the proneural wave front and act redundantly to regulate NB differentiation (Egger et al. 2007; Orihara-Ono et al. 2011; Yasugi et al. 2008). A is upregulated by E but downregulated by N, because AS-C expression is positively and negatively regulated by E and N, respectively (Fig. 5.2a, b). Secreted EGF ligands and Delta are produced only in undifferentiated NEs. Once the cells have differentiated into NBs, the cell type is completely switched and the cells start producing multiple types of neurons inside the brain (Fig. 5.1h) (Li et al. 2013; Suzuki et al. 2013). Because NB differentiation is an irreversible process, the state of differentiation must be maintained in differentiated NBs ($A = 1$). This maintenance was incorporated into the model by setting the rate of change in A to $e_a(1 - A)\max\{E - N, 0\}$. When N is greater than E, the value of $E - N$ is regarded as 0 to reflect a lack of dedifferentiation of NB cells (Fig. 5.2b).

AS-C triggers the expression of Dl (Kunisch et al. 1994). Loss of AS-C function downregulates EGF signaling activity (Sato et al. 2016). Thus, A upregulates E and D. We included the EGF and Delta production terms $a_e A(1 - A)$ and $a_d A(1 - A)$, respectively, to reflect that the production of these proteins is positively regulated by A when the cells are undifferentiated. These terms recapitulate the in vivo situation in which the EGF ligand and Delta are produced only in wave front cells but not in differentiated NBs (Fig. 5.1h).

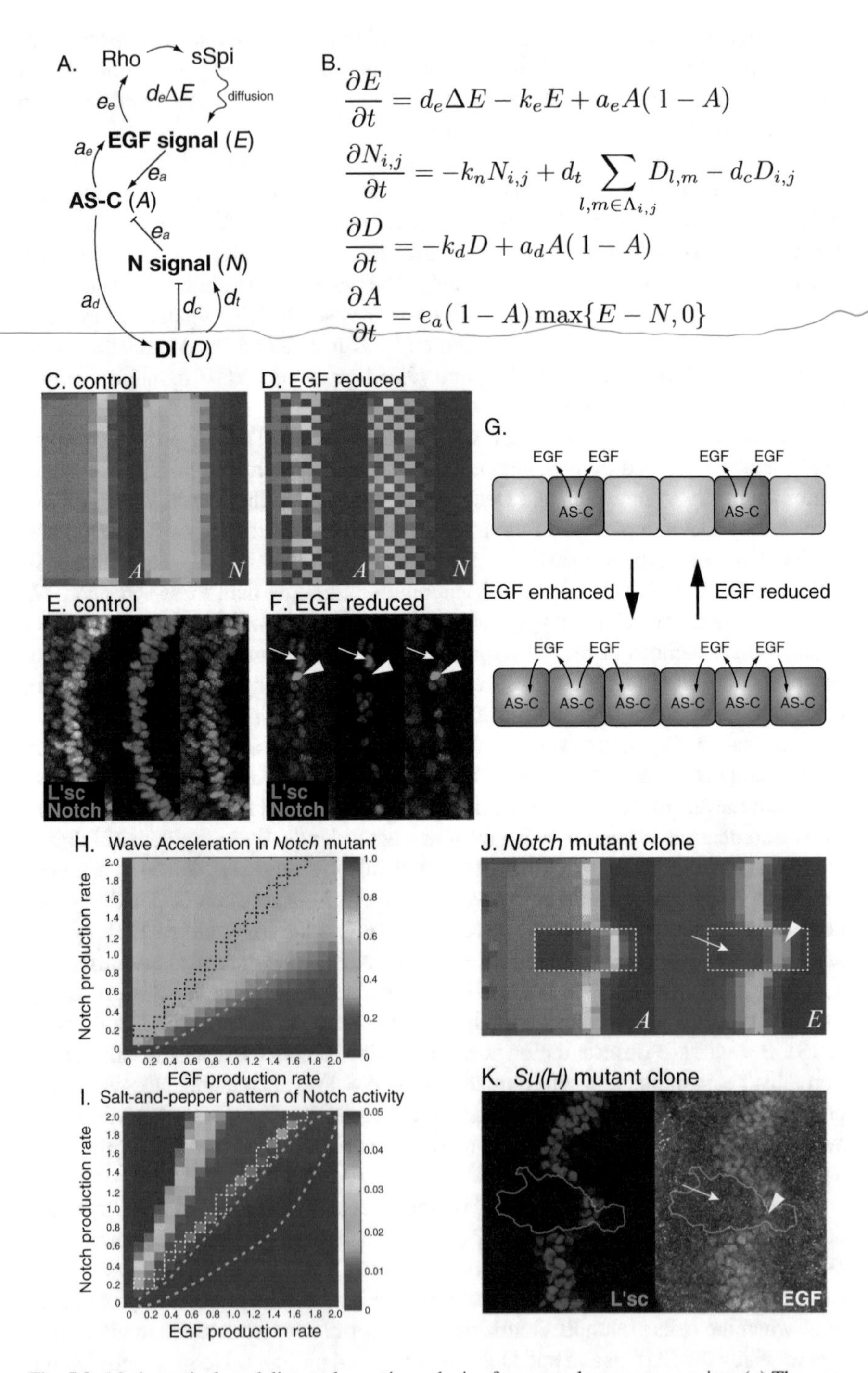

Fig. 5.2 Mathematical modeling and genetic analysis of proneural wave progression. (**a**) The gene regulatory network that includes EGF signaling (*E*), Notch signaling (*N*), Dl expression (*D*), and AS-C expression (*A*). (**b**) The 4-component model of the proneural wave. (**c, d**) Reducing the EGF

For simplicity, the terms for EGF autoregulation, EGF activation by Notch, and Delta upregulation by EGF were omitted from the original mathematical model (Sato et al. 2016). We have demonstrated that proneural wave progression in various mutant backgrounds can be reproduced even in the absence of these terms.

Phenomenological Versus Mechanistic Models

Mathematical modeling is an essential method with which to address complex systems that include multiple signaling systems. We can sometimes use a phenomenological model that recapitulates the behavior of a system without considering the detailed molecular mechanisms included in the system. Even if it is not based on real molecular mechanisms, such a phenomenological model provides fruitful biological suggestions. However, we can gain realistic biological insights only from a mechanistic model in which each one of the equations is based on a real molecular mechanism.

Although we tried to establish a mechanistic model of the proneural wave, it is virtually impossible to mechanistically model the process of cell differentiation, because differentiation is a very complex nonlinear process, which includes many genetic factors that induce differentiation and epigenetic factors that change the global state of the chromosomes. Nevertheless, it is widely accepted that the characteristics of cells are irreversibly and dramatically changed upon differentiation. We therefore assume that the expression level of AS-C is directly related to the state of differentiation, A, and that EGF (E) and Delta (D) increase only when NEs are being differentiated ($0 < A < 1$) (Sato et al. 2016). Indeed, a variable for the state of differentiation has been commonly used in other mathematical biological studies (Jorg et al. 2019; Corson et al. 2017).

In contrast to the case in the Jorg model, EGF (E) does not propagate on its own in our model. According to the biological observations discussed above, mutual regulation between EGF and AS-C is the core mechanism of proneural wave propagation (Fig. 5.1g).

Fig. 5.2 (continued) production rate, A and N show fluctuations in silico. (**e, f**) When EGF signaling is partially reduced in vivo, L'sc expression (*arrows*) and Notch activity (*arrowheads*) become stochastic. (**g**) AS-C expression becomes uniform and stochastic when EGF signaling is enhanced and reduced, respectively. (**h, i**) Phase diagrams of the 4-component model. Acceleration of wave progression in Notch mutant clones is observed in a wide range of parameter settings (**h**). Salt-and-pepper patterns are observed when the standard deviation of N is increased within the *white or black dotted lines. Orange dotted lines* indicate the area in which Notch signaling and EGF signaling are roughly equivalent. (**j**) Notch mutant clones are indicated by *white dotted lines*. E is activated when the proneural wave encounters Notch mutant cells (*arrowhead*), but is quickly inactivated (*arrow*). (**k**) EGF signaling (*white*) is transiently upregulated when the proneural wave encounters *Su(H)* mutant cells (*arrowhead*), but is eventually downregulated (*arrow*)

Our model is not stable when $A = 0$. Even a very small amount of noise added to E causes spontaneous differentiation that propagates the entire NE field (Tanaka et al. 2018). If we allow phenomenological modeling, it is possible to add an artificial term to the model to avoid such spontaneous differentiation. For example, a nonlinear reaction term, which has a small negative value when E has a small non-negative value, would make the model resistant to noise because small fluctuations in E suppress further increases in E (Jorg et al. 2019). This is a common strategy for adding noise resistance to a model. Instead of artificially manipulating the model, we added noise resistance in a mechanistic manner by including Jak/Stat, a signaling system that has been shown to suppress NB differentiation by augmenting Notch signaling (Fig. 5.1h, i) (Yasugi et al. 2008; Tanaka et al. 2018). Since Jak/Stat signaling constantly suppresses increases in A even in the presence of noise related to E, the model clearly becomes resistant to noise. Importantly, the noise-canceling effect of Jak/Stat has been experimentally reproduced by mildly reducing Stat activity through RNA interference (RNAi) (Tanaka et al. 2018). This is a good example of a mechanistic model revealing a new biological mechanism.

The Mathematical Model Reproduces the Essential Behaviors of the Proneural Wave

Our mathematical model reproduces the progression of the proneural wave in wild-type and various mutant conditions (Fig. 5.1i–k) (Sato et al. 2016). In the EGF mutant area, wave progression and NB differentiation disappear autonomously (Fig. 5.1j). In contrast, wave progression is accelerated in Notch and Delta mutant cells (Fig. 5.1k). Thus, the simple combination of EGF-mediated reaction diffusion and Notch-mediated lateral inhibition is sufficient to reproduce the essential behaviors of the proneural wave.

Although the model explicitly includes the lateral inhibition system, it reproduces wave propagation in the absence of a salt-and-pepper pattern (Fig. 5.2c). The lack of a salt-and-pepper pattern can be explained by the diffusible action of EGF. For simplicity, imagine that AS-C is activated in a small number of cells showing a salt-and-pepper pattern (Fig. 5.2g, top). In this context, EGF is produced from AS-C-positive cells and diffuses to neighboring undifferentiated cells. As a result, AS-C is upregulated in all NEs, obscuring the salt-and-pepper pattern (Fig. 5.2g, bottom).

If EGF counteracts salt-and-pepper pattern formation, reductions in EGF production should reproduce the formation of the salt-and-pepper pattern (Fig. 5.2g). This idea was tested by decreasing EGF production in the computer simulation, which resulted in salt-and-pepper-like fluctuations in AS-C and Notch (Fig. 5.2d). However, it will be important to test if this idea is true in vivo. Complete loss of EGF signaling eliminates proneural wave propagation (Fig. 5.1j) (Yasugi et al. 2010). We found a condition in which a partial reduction in EGF signaling caused salt-and-pepper-like fluctuations in L'sc expression and Notch activity

(Fig. 5.2e, f) (Sato et al. 2016). Although EGF signaling was uniformly reduced, the expression of L'sc and Notch activity became stochastic.

Theoretically, AS-C expression and Notch activity should become complementary as a result of the bistable nature of lateral inhibition (Fig. 5.1a). AS-C is upregulated in differentiating cells, while Notch is activated in adjacent undifferentiated cells. Importantly, this complementary pattern was also reproduced, as demonstrated by L'sc and Notch activity upon a partial reduction in EGF signaling (Fig. 5.2f, arrows and arrowheads), suggesting that Notch-mediated lateral inhibition is indeed implemented in the proneural wave in vivo.

Unfortunately, even in the presence of fluctuating L'sc expression, NBs were uniformly formed behind the proneural wave, as demonstrated by the expression of Dpn, an NB marker (Sato et al. 2016). As discussed earlier, NB differentiation is delayed but is eventually accomplished in the absence of AS-C activity (Yasugi et al. 2008). The final pattern of NB differentiation may be rescued by unidentified mechanisms.

Prior to the development of the mathematical model, it was not clear if Notch-mediated lateral inhibition was implemented in the proneural wave in vivo. As shown above, our model suggested that a simple combination of EGF diffusion and Notch-mediated lateral inhibition is sufficient to explain the diverse behavior of the proneural wave. Furthermore, our model predicted that a partial reduction in EGF signaling would cause the salt-and-pepper pattern. We validated this prediction by manipulating EGF signaling in real fly brains (Sato et al. 2016). Thus, molecular genetics studies that prove the mathematical prediction verify the validity of the mathematical model and provide new biological findings.

Two Distinct Functions of Notch Signaling

The above results indicate that Notch signaling has two distinct functions: regulation of wave progression speed and formation of the salt-and-pepper pattern. In the fly brain, the former is prominent but the latter is obscure. Extensive numerical simulations using more than 4000 parameter settings reproduced these two distinct outcomes (Fig. 5.2h, i) (Sato et al. 2016). Importantly, repression of AS-C expression was the only direct output of Notch signaling in our model (Fig. 5.2a). Nevertheless, the control of wave speed by Notch occurred over a wide range of parameter settings (Fig. 5.2h), while the salt-and-pepper pattern formation was restricted to a very narrow range of parameters (Fig. 5.2i). The result also indicates that the salt-and-pepper pattern is not formed when the magnitudes of EGF and Notch signals are roughly equivalent. This finding is consistent with in vivo observations that the salt-and-pepper pattern is hardly detectable in wild-type brains (Fig. 5.2e). Thus, we do not need to assume two distinct outputs of Notch signaling to explain its two distinct functions. Since the model is sufficient to reproduce the two distinct Notch functions, the mechanisms that differently control the dual Notch functions will be revealed by mathematical studies in the future.

Using the Model to Solve the Paradox of the Notch Mutant Phenotype

Proneural wave progression is accelerated in clones of cells in which Notch signaling is mutated (Fig. 5.1k) (Reddy et al. 2010; Yasugi et al. 2010). In addition, EGF signaling is also abolished in Notch mutant clones. On the other hand, EGF signaling is essential for proneural wave propagation in vivo (Fig. 5.1j) (Yasugi et al. 2010). If EGF signaling is lost in Notch mutant clones, the proneural wave should be eliminated because EGF is essential for wave progression. However, the proneural wave is accelerated in Notch and Delta mutant clones (Fig. 5.1k) (Reddy et al. 2010; Yasugi et al. 2010).

Results obtained with *Delta EGF* double mutant clones suggest that EGF signaling plays a key role in this paradoxical phenomenon. The wave is accelerated in *Delta* mutant clones. However, this phenotype is suppressed and wave progression is abolished in *Delta EGF* double mutant clones in vivo (as in Fig. 5.1j) (Yasugi et al. 2010). Simultaneous removal of Notch and EGF causes essentially the same result in numerical simulations (Fig. 5.2j) (Sato et al. 2016).

Therefore, we focused on the behavior of EGF signaling in the Notch mutant area *in silico*. Interestingly, the value of E was significantly elevated at the wave front within the Notch mutant area, but it was quickly diminished behind the wave front (Fig. 5.2j, arrow). Note that the value of E at the wave front in the Notch clone (shown by white dotted lines) was comparable to that in the control area (Fig. 5.2j, arrowhead). Thus, when Notch activity was instantaneously abolished in the Notch mutant area, the EGF activity was essentially unchanged at first. Because the time difference for A is positively regulated by the value of $E - N$ in our model (Fig. 5.2b), NB differentiation is accelerated if N tends to be zero, while E remains unchanged. The temporal increase in the value of $E - N$ at the wave front causes the wave acceleration (Fig. 5.2j).

It was important to confirm whether the above explanation was applicable to the in vivo situation. In *Su(H)* mutant clones, in which Notch signaling is eliminated, the proneural wave was accelerated, as demonstrated by visualization of L'sc expression (Fig. 5.2k, left). As reported previously, EGF signaling was abolished in the mutant clone behind the wave front (Fig. 5.2k, arrow). However, significant EGF signal activity as strong as that in neighboring control wave front cells was found at the mutant wave front (Fig. 5.2k, arrowhead). Thus, the mathematical model recapitulates proneural wave acceleration in Notch signaling mutant clones and clearly explains the hidden mechanism (Sato et al. 2016).

Expanding the Roles of Notch to Other Biological Phenomena

Notch signaling plays diverse roles when combined with other signaling systems (Doroquez and Rebay 2006; Dutt et al. 2004; Sundaram 2005; Kageyama et al. 2012; Kulesa et al. 2007). When these signaling systems form multiple feedback loops and demonstrate complex behaviors such as wave propagation or oscillation, mathematical modeling based on a mechanistic view point is essential to understand the core mechanism of the system behavior.

There are many examples of the interplay among multiple signaling systems during development. In the developing cerebral cortex, the oscillatory behavior of Notch signaling in neural progenitor cells is important for cell fate determination (Imayoshi et al. 2013). During this process, Notch signaling seems to regulate lateral inhibition. Additionally, the combined action of Notch and EGF maintains neural stem cells and neural progenitor cells (Aguirre et al. 2010). Similarly, vertebrate segmentation is controlled by the interplay between Notch and FGF signaling pathways (Kageyama et al. 2012). Notch signaling controls the synchronization between adjacent cells and regulates boundary formation. In contrast, Notch-mediated lateral inhibition causes desynchronization of NB differentiation between neighboring cells and negatively regulates NB formation during proneural wave progression (Sato et al. 2016). Thus, the roles of Notch signaling in the proneural wave appear to be the opposite of those observed in vertebrate segmentation.

The interplay between Notch and other signaling systems may also play important roles in cancer pathogenesis. The cross talk between Notch and EGF plays essential roles in lung and breast cancers (Pancewicz-Wojtkiewicz 2016; Baker et al. 2014). Understanding the cross talk of these pathways in cancer pathogenesis will be essential to improve and optimize cancer therapy in the future.

By using a mathematical model based on biological evidence, we found that a simple combination of Notch-mediated lateral inhibition and EGF-mediated reaction diffusion reproduced the complex behavior of the proneural wave. The key results predicted by the mathematical model were proven by molecular genetic experiments. Thus, mathematical modeling can be a powerful driving force for biological research. The use of similar interdisciplinary approaches may be essential to elucidate and target the core mechanisms of complex biological phenomena in animal development and cancer pathogenesis.

Acknowledgements We thank Masaharu Nagayama and Yoshitaro Tanaka for their critical comments. This work was supported by Core Research for Evolutional Science and Technology (CREST) from the Japan Science and Technology Agency (JST) (Grant JPMJCR14D3 to M.S.); Grants-in-Aid for Scientific Research on Innovative Areas and Grants-in-Aid for Scientific Research (B) and (C) from the Japanese Ministry of Education, Culture, Sports, Science and Technology (MEXT) (Grants JP17H05739, JP17H05761, JP17H03542, and JP19H04771 to M.S.; and Grants JP18H05099, 19K06674, and 19H04956 to T.Y.); Takeda Science Foundation (to M.S. and T.Y.); and a Grant for Cooperative Research on 'Network Joint Research Center for Materials and Devices' (to M.S.).

References

Aguirre A, Rubio ME, Gallo V (2010) Notch and EGFR pathway interaction regulates neural stem cell number and self-renewal. Nature 467(7313):323–327

Artavanis-Tsakonas S, Rand MD, Lake RJ (1999) Notch signaling: cell fate control and signal integration in development. Science 284(5415):770–776

Baker NE, Yu SY (1997) Proneural function of neurogenic genes in the developing *Drosophila* eye. Curr Biol 7(2):122–132

Baker AT, Zlobin A, Osipo C (2014) Notch-EGFR/HER2 bidirectional crosstalk in breast cancer. Front Oncol 4:360

Baonza A, Freeman M (2001) Notch signalling and the initiation of neural development in the *Drosophila* eye. Development 128(20):3889–3898

Cabrera CV, Alonso MC (1991) Transcriptional activation by heterodimers of the achaete-scute and daughterless gene products of *Drosophila*. EMBO J 10(10):2965–2973

Collier JR, Monk NA, Maini PK, Lewis JH (1996) Pattern formation by lateral inhibition with feedback: a mathematical model of delta-notch intercellular signalling. J Theor Biol 183(4):429–446

Corson F, Couturier L, Rouault H, Mazouni K, Schweisguth F (2017) Self-organized Notch dynamics generate stereotyped sensory organ patterns in *Drosophila*. Science 356(6337):eaai7407

del Alamo D, Rouault H, Schweisguth F (2011) Mechanism and significance of cis-inhibition in Notch signalling. Curr Biol 21(1):R40–R47

Doroquez DB, Rebay I (2006) Signal integration during development: mechanisms of EGFR and Notch pathway function and cross-talk. Crit Rev Biochem Mol Biol 41(6):339–385

Dutt A, Canevascini S, Froehli-Hoier E, Hajnal A (2004) EGF signal propagation during *C. elegans* vulval development mediated by ROM-1 rhomboid. PLoS Biol 2(11):e334

Egger B, Boone JQ, Stevens NR, Brand AH, Doe CQ (2007) Regulation of spindle orientation and neural stem cell fate in the *Drosophila* optic lobe. Neural Dev 2:1

Formosa-Jordan P, Ibanes M, Ares S, Frade JM (2012) Regulation of neuronal differentiation at the neurogenic wavefront. Development 139(13):2321–2329

Freeman M (1996) Reiterative use of the EGF receptor triggers differentiation of all cell types in the *Drosophila* eye. Cell 87(4):651–660

Ghysen A, Dambly-Chaudiere C, Jan LY, Jan YN (1993) Cell interactions and gene interactions in peripheral neurogenesis. Genes Dev 7(5):723–733

Heberlein U, Wolff T, Rubin GM (1993) The TGF beta homolog dpp and the segment polarity gene hedgehog are required for propagation of a morphogenetic wave in the *Drosophila* retina. Cell 75(5):913–926

Imayoshi I, Isomura A, Harima Y, Kawaguchi K, Kori H, Miyachi H, Fujiwara T, Ishidate F, Kageyama R (2013) Oscillatory control of factors determining multipotency and fate in mouse neural progenitors. Science 342(6163):1203–1208

Jarman AP, Grau Y, Jan LY, Jan YN (1993) atonal is a proneural gene that directs chordotonal organ formation in the *Drosophila* peripheral nervous system. Cell 73(7):1307–1321

Jarman AP, Grell EH, Ackerman L, Jan LY, Jan YN (1994) Atonal is the proneural gene for *Drosophila* photoreceptors. Nature 369(6479):398–400

Jorg DJ, Caygill EE, Hakes AE, Contreras EG, Brand AH, Simons BD (2019) The proneural wave in the *Drosophila* optic lobe is driven by an excitable reaction-diffusion mechanism. eLife 8:e40919

Kageyama R, Niwa Y, Isomura A, Gonzalez A, Harima Y (2012) Oscillatory gene expression and somitogenesis. Wiley Interdiscip Rev Dev Biol 1(5):629–641

Kawamori H, Tai M, Sato M, Yasugi T, Tabata T (2011) Fat/Hippo pathway regulates the progress of neural differentiation signaling in the *Drosophila* optic lobe. Dev Growth Differ 53(5):653–667

Kulesa PM, Schnell S, Rudloff S, Baker RE, Maini PK (2007) From segment to somite: segmentation to epithelialization analyzed within quantitative frameworks. Dev Dyn 236(6):1392–1402

Kunisch M, Haenlin M, Campos-Ortega JA (1994) Lateral inhibition mediated by the *Drosophila* neurogenic gene delta is enhanced by proneural proteins. Proc Natl Acad Sci U S A 91(21):10139–10143

Li X, Erclik T, Bertet C, Chen Z, Voutev R, Venkatesh S, Morante J, Celik A, Desplan C (2013) Temporal patterning of *Drosophila* medulla neuroblasts controls neural fates. Nature 498(7455):456–462

Lubensky DK, Pennington MW, Shraiman BI, Baker NE (2011) A dynamical model of ommatidial crystal formation. Proc Natl Acad Sci U S A 108(27):11145–11150

Ma C, Zhou Y, Beachy PA, Moses K (1993) The segment polarity gene hedgehog is required for progression of the morphogenetic furrow in the developing *Drosophila* eye. Cell 75(5):927–938
Neumann CJ, Nuesslein-Volhard C (2000) Patterning of the zebrafish retina by a wave of sonic hedgehog activity. Science 289(5487):2137–2139
Orihara-Ono M, Toriya M, Nakao K, Okano H (2011) Downregulation of Notch mediates the seamless transition of individual *Drosophila* neuroepithelial progenitors into optic medullar neuroblasts during prolonged G1. Dev Biol 351(1):163–175
Pancewicz-Wojtkiewicz J (2016) Epidermal growth factor receptor and notch signaling in non-small-cell lung cancer. Cancer Med 5(12):3572–3578
Pennington MW, Lubensky DK (2010) Switch and template pattern formation in a discrete reaction-diffusion system inspired by the *Drosophila* eye. Eur Phys J E Soft Matter 33(2):129–148
Reddy BV, Rauskolb C, Irvine KD (2010) Influence of fat-hippo and notch signaling on the proliferation and differentiation of *Drosophila* optic neuroepithelia. Development 137(14):2397–2408
Sato M, Suzuki T, Nakai Y (2013) Waves of differentiation in the fly visual system. Dev Biol 380(1):1–11
Sato M, Yasugi T, Minami Y, Miura T, Nagayama M (2016) Notch-mediated lateral inhibition regulates proneural wave propagation when combined with EGF-mediated reaction diffusion. Proc Natl Acad Sci U S A 113(35):E5153–E5162
Simpson P (1990) Lateral inhibition and the development of the sensory bristles of the adult peripheral nervous system of *Drosophila*. Development 109(3):509–519
Sprinzak D, Lakhanpal A, Lebon L, Santat LA, Fontes ME, Anderson GA, Garcia-Ojalvo J, Elowitz MB (2010) Cis-interactions between Notch and Delta generate mutually exclusive signalling states. Nature 465(7294):86–90
Sundaram MV (2005) The love–hate relationship between Ras and Notch. Genes Dev 19(16):1825–1839
Suzuki T, Kaido M, Takayama R, Sato M (2013) A temporal mechanism that produces neuronal diversity in the *Drosophila* visual center. Dev Biol 380(1):12–24
Tanaka Y, Yasugi T, Nagayama M, Sato M, Ei SI (2018) JAK/STAT guarantees robust neural stem cell differentiation by shutting off biological noise. Sci Rep 8(1):12484
Urban S, Lee JR, Freeman M (2001) *Drosophila* rhomboid-1 defines a family of putative intramembrane serine proteases. Cell 107(2):173–182
Yang HJ, Silva AO, Koyano-Nakagawa N, McLoon SC (2009) Progenitor cell maturation in the developing vertebrate retina. Dev Dyn 238(11):2823–2836
Yasugi T, Umetsu D, Murakami S, Sato M, Tabata T (2008) *Drosophila* optic lobe neuroblasts triggered by a wave of proneural gene expression that is negatively regulated by JAK/STAT. Development 135:1471–1480
Yasugi T, Sugie A, Umetsu D, Tabata T (2010) Coordinated sequential action of EGFR and Notch signaling pathways regulates proneural wave progression in the *Drosophila* optic lobe. Development 137(19):3193–3203
Yuan Z, Praxenthaler H, Tabaja N, Torella R, Preiss A, Maier D, Kovall RA (2016) Structure and function of the Su(H)-Hairless repressor complex, the major antagonist of Notch signaling in *Drosophila melanogaster*. PLoS Biol 14(7):e1002509
zur Lage PI, Prentice DR, Holohan EE, Jarman AP (2003) The *Drosophila* proneural gene amos promotes olfactory sensillum formation and suppresses bristle formation. Development 130(19):4683–4693

Chapter 6
A Nucleolar Protein, Nepro, Is Essential for the Maintenance of Early Neural Stem Cells and Preimplantation Embryos

Tetsuichiro Saito

Abstract Notch signaling is required for maintaining neural stem cells (NSCs) in the developing brain. NSCs have potential to give rise to many neuronal types in the early telencephalon, and the potential decreases as embryonic development proceeds. *Nepro*, which encodes a unique nucleolar protein and is activated downstream of Notch, is essential for maintaining NSCs in the early telencephalon. *Nepro* is also expressed at basal levels and required for maintaining the preimplantation embryo, by repressing mitochondria-associated p53 apoptotic signaling. Notch signaling also controls dendritic complexity in mitral cells, major projection neurons in the olfactory bulb, showing that many steps of neural development involve Notch signaling.

Keywords Mib1 · Maml1 · Nepro · Hes1 · Neural stem cell · Radial glia · Neuron · Dendrite · Neocortex · Telencephalon · Blastocyst · Morula

Development of the Neocortex

NSCs are defined as cells that are capable to self-renew and give rise to both glial cells and at least one type of neurons. The mammalian neocortex is the center of higher cognitive functions and contains many types of neurons in a six-layered structure (Tasic et al. 2016, 2018; Saunders et al. 2018), which differentiate from spatially and temporally distinct populations of NSCs. The vast majority of neurons in the neocortex are projection neurons, which are excitatory and glutamatergic. They are produced from NSCs in the ventricular zone (VZ) of the dorsal telencephalon and migrate radially. The other neurons are GABAergic interneurons,

T. Saito (✉)
Department of Developmental Biology, Graduate School of Medicine, Chiba University, Chiba, Japan
e-mail: tesaito@faculty.chiba-u.jp

J. Reichrath, S. Reichrath (eds.), *Notch Signaling in Embryology and Cancer*, Advances in Experimental Medicine and Biology 1218,
https://doi.org/10.1007/978-3-030-34436-8_6

which are generated from NSCs in the ventral telencephalon and migrate tangentially (Lodato and Arlotta 2015).

At the beginning of telencephalon development, neuroepithelial cells divide symmetrically to increase their numbers. At around mouse embryonic (E) 9.5, they convert to radial glial cells, which function as NSCs at early stages. In the dorsal telencephalon, each NSC divides asymmetrically to give rise to a NSC and a neuron or a basal progenitor, which divides symmetrically to form two neurons. Those neurons migrate into the cortical plate (CP) using radial glial fibers as scaffolds. Early-born neurons occupy low layers of the CP, and later-born neurons migrate into more superficial layers so that layers are sequentially formed from the layer VI to layer II. NSCs change their potential as embryonic development proceeds. Early NSCs are competent to generate neurons of all layers, and NSCs gradually lose the competence to generate neurons of lower layers (McConnell 1988; Frantz and McConnell 1996; Mizutani and Saito 2005). A subpopulation of NSCs become quiescent between E13.5 and E15.5 and are reactivated at postnatal stages to give rise to adult NSCs, which continue to produce neurons in restricted areas such as the hippocampus throughout life (Fuentealba et al. 2015; Furutachi et al. 2015). RGCs at later stages, which are no longer NSCs, produce only glial cells such as astrocytes (Dwyer et al. 2016). Therefore, the maintenance of NSCs is essential for generating the diversity of many cell types and for the function of the neocortex.

Notch Signaling Is Required for the Maintenance of NSCs

Notch signaling plays a pivotal role in the maintenance of NSCs (Homem et al. 2015). Misexpression of a constitutive active form of *Notch* (*caNotch*) inhibits neuronal differentiation and maintains NSCs in the dorsal telencephalon (Gaiano et al. 2000; Saito and Nakatsuji 2001). *caNotch*-misexpressing NSCs continue cell division in the VZ and resume generating neurons after switching off *caNotch* (Mizutani and Saito 2005). The neurons generated from the NSCs that did not produce early-born neurons by *caNotch* show the same characters of later-born neurons, indicating that the potential of NSCs decreases with embryonic development, despite producing no neurons (Mizutani and Saito 2005).

NSCs receive ligands such as Dll1 from basal progenitors and neurons, leading to the cleavage of the intracellular domains of Notch receptors, which enter the nucleus, form a complex with Rbpj and a coactivator Mastermind-like (Maml), and activate transcription of target genes, such as *Hes1* (Guruharsha et al. 2012). Hes1 represses proneural genes, such as *Ascl1*, thereby inhibiting neurogenesis and maintaining NSCs. Mindbomb homolog 1 (Mib1), which ubiquitinates the intracellular domains of Notch ligands, is required for their function (Bray 2016).

At an early stage, NSCs themselves express Dll1 in an oscillatory manner, which activates Notch signaling in neighboring NSCs for their maintenance (Shimojo et al. 2016).

Both Nepro and Hes Are Required for the Maintenance of Early NSCs

Nepro is expressed at high levels in NSCs of the brain only at early stages (E9.5 to E12.5) and activated downstream of Notch (Muroyama and Saito 2009; Saito 2012). Misexpression of *caNotch* inhibits neuronal differentiation, thereby keeping transfected enhanced yellow fluorescent protein (EYFP)-positive cells as NSCs in the VZ (Fig. 6.1a). Cotransfection of a *Nepro*-specific siRNA or a dominant negative form of *Nepro* (*dnNepro*) overrides the effect of caNotch, causing differentiation of many neurons in the CP (Fig. 6.1b, c). This finding indicates that Nepro is required for the maintenance of NSCs downstream of Notch.

On the other hand, Nepro is not sufficient for maintaining NPCs in the absence of Notch signaling. A dominant negative form of *MAML1* (*dnMAML1*) blocks Notch signaling, leading to precocious differentiation of NSCs into neurons in the CP (Fig. 6.2b), compared with the transfection of *EYFP* alone as a control, which labels both NSCs in the VZ and differentiated neurons in the CP (Fig. 6.2a).

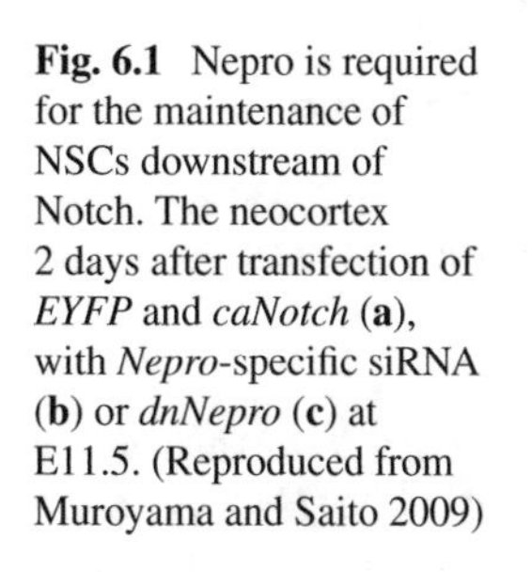

Fig. 6.1 Nepro is required for the maintenance of NSCs downstream of Notch. The neocortex 2 days after transfection of *EYFP* and *caNotch* (**a**), with *Nepro*-specific siRNA (**b**) or *dnNepro* (**c**) at E11.5. (Reproduced from Muroyama and Saito 2009)

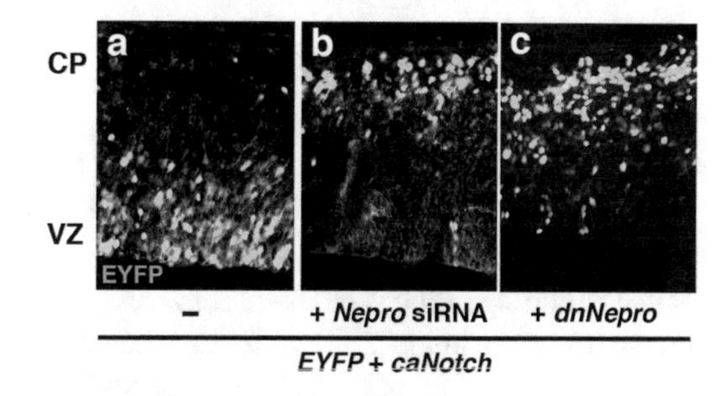

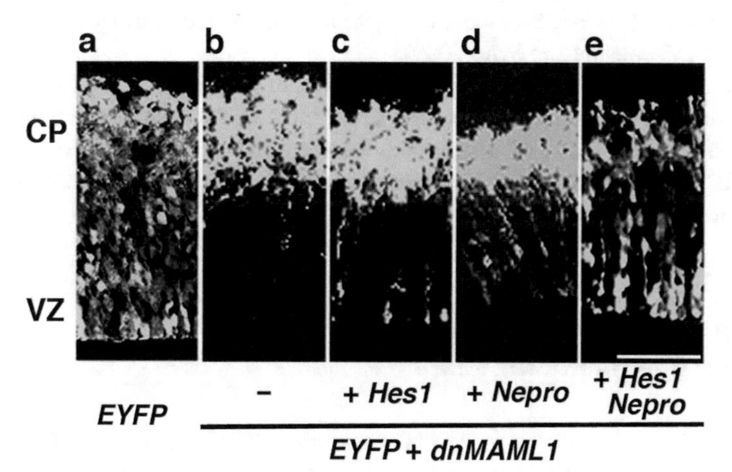

Fig. 6.2 Both Hes and Nepro are essential for the maintenance of early NSCs. The neocortex 2 days after transfection of *EYFP* as a control (**a**) and cotransfection of *EYFP* and *dnMAML1* (**b**), with *Hes1* (**c**), or *Nepro* (**d**), *Hes1* plus *Nepro* (**e**) at E11.5. Scale bar: 50 μm. (Reproduced from Muroyama and Saito 2009)

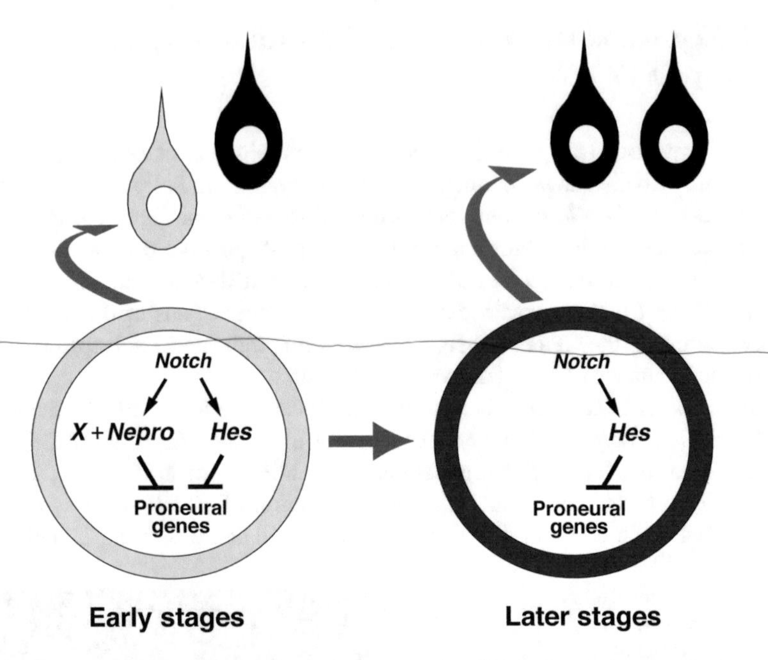

Fig. 6.3 Schematic illustration of the cascades to maintain NSCs in developing dorsal telencephalon at early stages and stages later than E15.5. Early NSCs give rise to neurons of all layers in the neocortex, whereas E15.5 NSCs generate only neurons of upper layers

When Notch signaling is blocked, cotransfection of both *Hes1* and *Nepro* efficiently maintains NSCs in the VZ (Fig. 6.2e), whereas transfection of either *Hes1* or *Nepro* is not sufficient (Fig. 6.2c, d). Similar phenotypes are observed when Notch signaling is blocked by a γ-secretase inhibitor, which inhibits the cleavage and activation of Notch (Muroyama and Saito 2009).

Overexpression of *Nepro* inhibits expression of proneural genes, such as *Ascl1* and *Neurog2*, and neurogenesis at early stages but not later stages (later than E15.5), whereas *caNotch* and *Hes1* inhibit both the early and later stages (Muroyama and Saito 2009), suggesting that Nepro requires another unknown early factor (X) for its function (Fig. 6.3).

Nepro Is also Required for the Maintenance of the Preimplantation Embryo

Whereas *Nepro*$^{+/-}$ mice grow normally and are fertile, *Nepro*$^{-/-}$ embryos do not form blastocysts, causing apoptotic cell death at the morula stage (Hashimoto et al. 2015). In *Nepro*$^{-/-}$ embryos, cytochrome *c* release and caspase-3 cleavage are activated concomitant with the increase of mitochondria-associated p53, suggesting that mitochondria-associated p53 apoptotic signaling is repressed downstream of Nepro in the preimplantation embryo (Fig. 6.4).

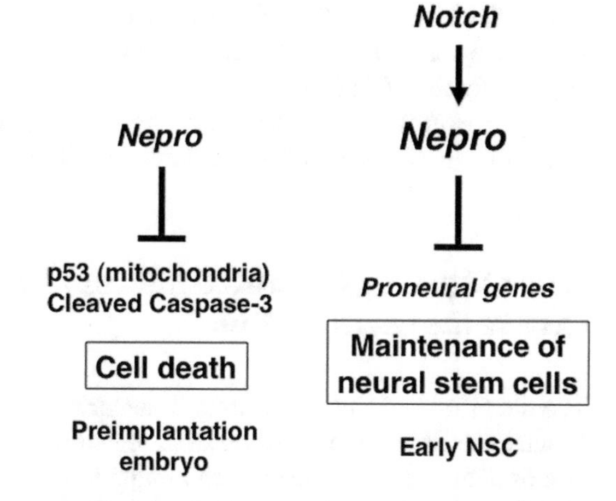

Fig. 6.4 Schematic illustration of Nepro cascades in the preimplantation embryo and early NSCs

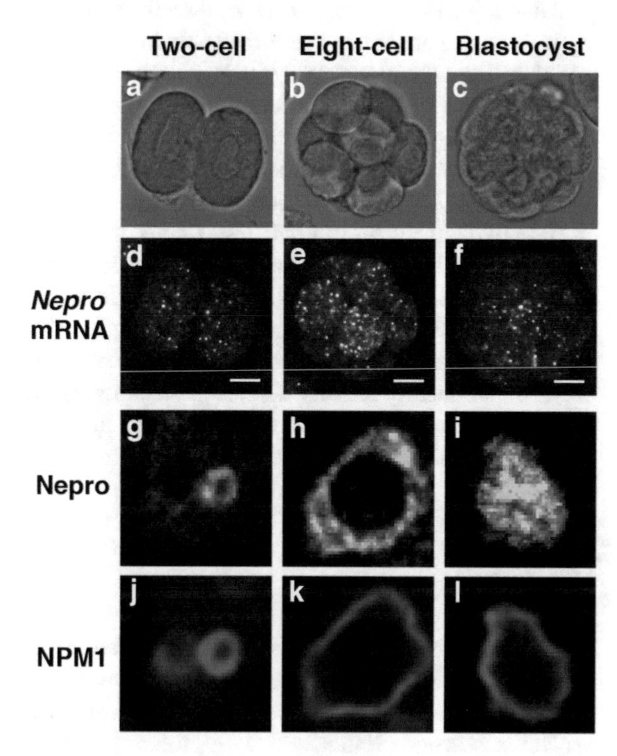

Fig. 6.5 Nepro expression and localization at preimplantation stages. Images of differential interference contrast (**a**–**c**), fluorescent in situ hybridization of *Nepro* (**d**–**f**) and immunostaining of Nepro (**g**–**i**) and NPM1 (**j**–**l**) in the nucleus at the two-cell (left), eight-cell (middle), and blastocyst (right) stages. Scale bars: 20 μm. (Reproduced from Hashimoto et al. 2015)

Nepro is expressed, albeit at low levels, from the two-cell to blastocyst stages in the preimplantation embryo (Fig. 6.5a–f). The Nepro protein is localized in nucleolus precursor bodies (NPBs) or nucleoli, which are labeled with a typical nucleolar protein NPM1 (alias nucleophosmin 1) (Fig. 6.5g–l).

The size of NPBs and nucleoli is smaller, and their number per nucleus is increased in *Nepro*$^{-/-}$ embryos from the four-cell to morula stages, suggesting that the formation of the NPB and nucleolus is abnormal in *Nepro*$^{-/-}$ embryos. Moreover, 18S rRNA and rpS6, which are well-studied components of the ribosome, are greatly diminished in the cytoplasm of *Nepro*$^{-/-}$ embryos at the morula stage, suggesting that Nepro is required for the function of the nucleolus.

Nepro expression is not affected by *dnMAML1* in preimplantation embryos, suggesting that *Nepro* is not downstream of Notch at preimplantation stages.

Notch Signaling Controls Dendritic Complexity of Mitral Cells in the Olfactory Bulb

The direct role of Notch signaling in dendrites of neurons was not clear, because Notch controls many steps from NSCs to neuronal differentiation, including inhibition of neuronal differentiation and migration of postmitotic neurons along radial glial fibers (Pierfelice et al. 2011). To directly examine the function of Notch signaling in dendrites, it has been perturbed only in mitral cells that finished migration and settled in the olfactory bulb, by using a conditional misexpression method, combined with in vivo electroporation (Saito and Nakatsuji 2001; Saito 2006) and the Tet-controlled gene expression system (Sato et al. 2013). Whereas misexpression of *caNotch* reduces their dendritic complexity, *dnMAML1* increases, suggesting that dendritic complexity is controlled by Notch signaling in mitral cells (Muroyama et al. 2016). In vitro cultured mitral cells reduce dendritic complexity by the treatment of a Notch ligand Jag1. Moreover, dendritic complexity is affected by a dosage of *Maml1*. Many *Maml1*$^{+/-}$ pups are defective in homing behavior, which is implicated in olfactory information processing. Furthermore, dendritic complexity is increased by inhibiting the activity of Notch ligands with a dominant negative form of Mib1 in olfactory sensory neurons, indicating that olfactory sensory axons activate the Notch pathway and control dendrites in their target postsynaptic neurons, mitral cells (Muroyama et al. 2016).

Discussion

Many steps of neural development are controlled by Notch signaling. Its functional analyses should be spatially and temporally well designed to dissect individual steps. There are contradictory reports for Notch signaling in dendritic complexity of neocortex neurons (Sestan et al. 1999; Redmond et al. 2000; Breunig et al. 2007), probably because Notch signaling was not examined specifically in a particular step or cell type of interest, despite many neuronal types. The role of Notch signaling in dendrites has been clarified, by focusing on mitral cells in the olfactory bulb, which has a simpler structure and fewer neuronal types than the neocortex. It remains to be determined whether Notch-controlled dendritic complexity is specific to mitral cells or common to other neuronal types. Dendritic morphologies of neurons are important for information processing and characteristic of each neuronal type. The control of mitral cell dendritic complexity by olfactory sensory axons is the first finding that axons initiate Notch signaling in their postsynaptic dendrites. Many molecules of axons, such as neurotransmitters and neurotrophins, affect dendrites (Lefebvre et al. 2015). It will be important to learn how these molecules interact with Notch signaling.

Nepro is necessary for two types of multipotent stem cells: blastomeres and early NSCs. Another nucleolar protein, GNL3 (alias nucleostemin), which is a GTP binding protein, is highly expressed in many types of stem cells and cancer cells and is required for their proliferation (Tsai 2014). *GNL3*$^{-/-}$ embryos do not form blastocysts, similar to *Nepro*$^{-/-}$. The function of GNL3, however, will be different from that of Nepro in the preimplantation embryo, because GNL3 is involved in genome maintenance in self-renewal of cells (Tsai 2014). GNL3 homologs are found in not only animals but also plants and fungi, in contrast to no invertebrate homolog of Nepro. Each vertebrate has a single Nepro homolog.

The temporal change of NSC's potential is a key to generate cellular diversity in the neocortex. Mechanisms underlying the change are still poorly understood. The expression and function of Nepro divide the states of NSCs at least into two: Nepro-dependent early and Nepro-independent later ones (Fig. 6.3). It remains to be determined whether Nepro is involved in multipotency of early NSCs.

In contrast to the preimplantation embryo, mitochondria-associated p53 apoptotic signaling appears not to be activated by the *Nepro*-specific siRNA or *dnNepro* in early NSCs, suggesting that cascades downstream of Nepro differ between the preimplantation embryo and early NSCs (Fig. 6.4). Nepro is expressed at lower levels in the preimplantation embryo than early NSCs, suggesting that *Nepro* is expressed at basal levels in the preimplantation embryo and activated by Notch signaling in early NSCs.

Conclusion

Notch signaling controls a series of events from the onset of neuronal differentiation to dendrite morphogenesis. A Notch effector Nepro is essential for the maintenance of early NSCs and preimplantation embryos. The function and regulation of Nepro differ between those two types of multipotent stem cells.

Acknowledgments This work was supported by JSPS KAKENHI Grant Number 18K19369.

References

Bray SJ (2016) Notch signalling in context. Nat Rev Mol Cell Biol 17:722–735. https://doi.org/10.1038/nrm.2016.94

Breunig JJ, Silbereis J, Vaccarino FM et al (2007) Notch regulates cell fate and dendrite morphology of newborn neurons in the postnatal dentate gyrus. Proc Natl Acad Sci U S A 104:20558–20563. https://doi.org/10.1073/pnas.0710156104

Dwyer ND, Chen B, Chou SJ et al (2016) Neural stem cells to cerebral cortex: emerging mechanisms regulating progenitor behavior and productivity. J Neurosci 36:11394–11401. https://doi.org/10.1523/JNEUROSCI.2359-16.2016

Frantz GD, McConnell SK (1996) Restriction of late cerebral cortical progenitors to an upper-layer fate. Neuron 17:55–61. https://doi.org/10.1016/S0896-6273(00)80280-9

Fuentealba LC, Rompani SB, Parraguez JI et al (2015) Embryonic origin of postnatal neural stem cells. Cell 161:1644–1655. https://doi.org/10.1016/j.cell.2015.05.041

Furutachi S, Miya H, Watanabe T et al (2015) Slowly dividing neural progenitors are an embryonic origin of adult neural stem cells. Nat Neurosci 18:657–665. https://doi.org/10.1038/nn.3989

Gaiano N, Nye JS, Fishell G (2000) Radial glial identity is promoted by Notch1 signaling in the murine telencephalon. Neuron 26:395–404. https://doi.org/10.1016/S0896-6273(00)81172-1

Guruharsha KG, Kankel MW, Artavanis-Tsakonas S (2012) The Notch signalling system: recent insights into the complexity of a conserved pathway. Nat Rev Genet 13:654–666. https://doi.org/10.1038/nrg3272

Hashimoto M, Sato T, Muroyama Y et al (2015) Nepro is localized in the nucleolus and essential for preimplantation development in mice. Dev Growth Differ 57:529–538. https://doi.org/10.1111/dgd.12232

Homem CC, Repic M, Knoblich JA (2015) Proliferation control in neural stem and progenitor cells. Nat Rev Neurosci 16:647–659. https://doi.org/10.1038/nrn4021

Lefebvre JL, Sanes JR, Kay JN (2015) Development of dendritic form and function. Annu Rev Cell Dev Biol 31:741–777. https://doi.org/10.1146/annurev-cellbio-100913-013020

Lodato S, Arlotta P (2015) Generating neuronal diversity in the mammalian cerebral cortex. Annu Rev Cell Dev Biol 31:699–720. https://doi.org/10.1146/annurev-cellbio-100814–125353

McConnell SK (1988) Fates of visual cortical neurons in the ferret after isochronic and heterochronic transplantation. J Neurosci 8:945–974. https://doi.org/10.1523/JNEUROSCI.08-03-00945.1988

Mizutani K, Saito T (2005) Progenitors resume generating neurons after temporary inhibition of neurogenesis by Notch activation in the mammalian cerebral cortex. Development 132:1295–1304. https://doi.org/10.1242/dev.01693

Muroyama Y, Saito T (2009) Identification of *Nepro*, a gene required for the maintenance of neocortex neural progenitor cells downstream of Notch. Development 136:3889–3893. https://doi.org/10.1242/dev.039180

Muroyama Y, Baba A, Kitagawa M et al (2016) Olfactory sensory neurons control dendritic complexity of mitral cells via Notch signaling. PLoS Genet 12:e1006514. https://doi.org/10.1371/journal.pgen.1006514

Pierfelice T, Alberi L, Gaiano N (2011) Notch in the vertebrate nervous system: an old dog with new tricks. Neuron 69:840–855. https://doi.org/10.1016/j.neuron.2011.02.031

Redmond L, Oh SR, Hicks C et al (2000) Nuclear Notch1 signaling and the regulation of dendritic development. Nat Neurosci 3:30–40. https://doi.org/10.1038/71104

Saito T (2006) In vivo electroporation in the embryonic mouse central nervous system. Nat Protoc 1:1552–1558. https://www.nature.com/articles/nprot.2006.276

Saito T (2012) NEPRO: a novel Notch effector for maintenance of neural progenitor cells in the neocortex. Adv Exp Med Biol 727:61–70. https://doi.org/10.1007/978-1-4614-0899-4_5

Saito T, Nakatsuji N (2001) Efficient gene transfer into the embryonic mouse brain using in vivo electroporation. Dev Biol 240:237–246. https://doi.org/10.1006/dbio.2001.0439

Sato T, Muroyama Y, Saito T (2013) Inducible gene expression in postmitotic neurons by an in vivo electroporation-based tetracycline system. J Neurosci Methods 214:170–176. https://doi.org/10.1016/j.jneumeth.2013.01.014

Saunders A, Macosko EZ, Wysoker A et al (2018) Molecular diversity and specializations among the cells of the adult mouse brain. Cell 174:1015–1030. https://doi.org/10.1016/j.cell.2018.07.028

Sestan N, Artavanis-Tsakonas S, Rakic P (1999) Contact-dependent inhibition of cortical neurite growth mediated by notch signaling. Science 286:741–746. https://doi.org/10.1126/science.286.5440.741

Shimojo H, Isomura A, Ohtsuka T et al (2016) Oscillatory control of Delta-like1 in cell interactions regulates dynamic gene expression and tissue morphogenesis. Genes Dev 30:102–116. https://doi.org/10.1101/gad.270785.115

Tasic B, Menon V, Nguyen TN et al (2016) Adult mouse cortical cell taxonomy revealed by single cell transcriptomics. Nat Neurosci 19:335–346. https://doi.org/10.1038/nn.4216

Tasic B, Yao Z, Graybuck LT et al (2018) Shared and distinct transcriptomic cell types across neocortical areas. Nature 563:72–78. https://doi.org/10.1038/s41586-018-0654-5

Tsai RYL (2014) Turning a new page on nucleostemin and self-renewal. J Cell Sci 127:3885–3891. https://doi.org/10.1242/jcs.154054

Chapter 7
Role of Notch Signaling in Leg Development in *Drosophila melanogaster*

Sergio Córdoba and Carlos Estella

Abstract Notch pathway plays diverse and fundamental roles during animal development. One of the most relevant, which arises directly from its unique mode of activation, is the specification of cell fates and tissue boundaries. The development of the leg of *Drosophila melanogaster* is a fine example of this Notch function, as it is required to specify the fate of the cells that will eventually form the leg joints, the flexible structures that separate the different segments of the adult leg. Notch activity is accurately activated and maintained at the distal end of each segment in response to the proximo-distal patterning gene network of the developing leg. Region-specific downstream targets of Notch in turn regulate the formation of the different types of joints. We discuss recent findings that shed light on the molecular and cellular mechanisms that are ultimately governed by Notch to achieve epithelial fold and joint morphogenesis. Finally, we briefly summarize the role that Notch plays in inducing the nonautonomous growth of the leg. Overall, this book chapter aims to highlight leg development as a useful model to study how patterning information is translated into specific cell behaviors that shape the final form of an adult organ.

Keywords Notch · Drosophila · Leg · Pattern formation · Morphogenesis · Joint · Rho1 · Dysf · Rho GTPases · Apical constriction · Apoptosis · Myo II

S. Córdoba (✉)
Centro de Biología Molecular "Severo Ochoa", CSIC-UAM, Madrid, Spain

Department of Biology, New York University, New York, NY, USA
e-mail: sc7880@nyu.edu

C. Estella (✉)
Centro de Biología Molecular "Severo Ochoa", CSIC-UAM, Madrid, Spain
e-mail: cestella@cbm.csic.es

J. Reichrath, S. Reichrath (eds.), *Notch Signaling in Embryology and Cancer*, Advances in Experimental Medicine and Biology 1218,
https://doi.org/10.1007/978-3-030-34436-8_7

Introduction

The first description of a Notch phenotype was provided in 1914 by John S. Dexter, after screening mutant flies that presented notches in their wing margins, and some years later, in 1917, Thomas H. Morgan isolated the first Notch allele (Dexter 1914; Morgan 1917). More than 100 years later, we have reached a detailed knowledge of the molecular components of the pathway and the Notch influence in development and disease (Bray 2006; Hori et al. 2013; Penton et al. 2012; Bigas and Espinosa 2018). The Notch pathway is evolutionarily conserved and plays multiple roles during animal development. Notch pathway activation mechanism is unique; it relies in direct cell-cell contact between the sender and the receiving cells instead of depending on a secreted signal, as both Notch receptor and its ligands are transmembrane proteins. The efficient binding of Notch receptor to its ligand in an adjacent cell causes the proteolytic cleavage of Notch which then translocates into the nucleus to regulate target gene expression. Despite its simple mode of activation, the Notch pathway is essential for a wide range of developmental processes, ranging from binary cell-fate decisions to stem cell renewal (Bray 1998; Andersson et al. 2011). Thus, the capability of Notch to direct different developmental processes does not arise from the simple molecular design of the core pathway, but rather the transcriptional outcome of Notch activation is dependent on the cellular context and signaling dynamics (Bray and Bernard 2010; Henrique and Schweisguth 2019).

The molecular details of the Notch pathway have been extensively studied and are the subject of many specialized reviews (Bray 2006; Henrique and Schweisguth 2019; Kopan and Ilagan 2009); therefore only a shallow explanation is provided in this work. In brief, ligand-receptor interaction triggers the proteolytic cleavage of the Notch receptor and the release of its intracellular domain (Notch^{ICD}) that enters the nucleus to regulate gene transcription. Both ligands and receptor are transmembrane proteins that suffer complex post-translational modifications that modulate the efficiency of ligand-receptor interaction, the availability of the proteins at the cell surface, as well as the signaling outcome (Bigas and Espinosa 2018; Kopan and Ilagan 2009; Panin et al. 1997; Okajima and Irvine 2002; Harvey and Haltiwanger 2018).

After Notch receptor cleavage, Notch^{ICD} translocates to the nucleus where it interacts with a family of proteins known as CSL (named after their initials: CBF1/RBPJ-k in mammals, Su(H) in *Drosophila*, and Lag1 in *Caenorhabditis elegans*) which confers DNA-binding specificity to the complex (Kopan and Ilagan 2009; Greenwald 2012). When the Notch pathway is not activated, CSL proteins recognize their specific binding sites at the regulatory region of Notch target genes and recruit corepressors (Co-R) to inhibit gene transcription, which is known as *default repression* (Barolo et al. 2002). Upon Notch activation, Notch^{ICD} binds to CSL, displaces Co-R, and allows the direct regulation of target genes. Therefore, Notch can regulate transcription in a *permissive* manner, simply alleviating an existent repression. In addition, Notch^{ICD} can also play an *inductive* role, forming a tertiary complex with CSL proteins and coactivators to enhance the expression of target genes (Kopan and Ilagan 2009; Bray and Furriols 2001; Lai 2002). Additionally, the presence of other tissue-specific transcription factors and the crosstalk with other signaling pathways are usually necessary to enhance target gene transcription, as Su(H) sites alone are poor mediating transcriptional activation. This requirement allows differential signaling outcomes upon Notch activation. Therefore, the same transduction mechanism could elicit different responses in different tissues, depending on the preexistent cellular context (reviewed in Bray and Furriols (2001)).

Notch pathway is remarkably conserved among metazoans, and its core components are very similar from the worm *C. elegans* to mammals (Kopan and Ilagan 2009). The study of the Notch pathway in model organisms such as *C. elegans* and *Drosophila* has provided in the past some of the most important breakthroughs in the study of Notch mechanism and function (Greenwald 2012). Using *Drosophila* as a model organism presents several advantages in studying Notch function, most importantly the lack of redundancy of the pathway components. There is only one Notch receptor in *Drosophila*, whereas four paralogs are present in mammals (NOTCH1-4), and two ligands, Delta (Dl), related to mammalian Delta-like (Dll1, Dll3, and Dll4), and Serrate (Ser), ortholog of Jagged1 and Jagged2 (Kopan and Ilagan 2009). Additionally, *Drosophila* presents a wide variety of genetic tools that allow precise analysis and manipulation (del Valle Rodriguez et al. 2012).

In this chapter, we summarize the regulatory role of Notch pathway during leg development in *Drosophila*. In this context, Notch activation is required, at least, for two different processes. First, its restricted activity in discrete bands of cells is necessary for the formation of the joints that separate adult leg segments and allow appendage articulation (Fig. 7.1). And second, Notch activity from these borders acts as a regulator of leg growth in a nonautonomous manner. Here, we review the existent knowledge regarding how Notch signaling is spatially localized in response to proximo-distal leg patterning and the molecular and cellular mechanisms downstream of Notch activity that direct joint formation and leg growth.

Leg Segmentation and Positioning of Notch Activity

Leg Specification and Proximo-distal Patterning

Appendages are structures that project out from an animal's body wall, including legs, antennae, wings, or genitalia, and their presence allows the implementation of different biological functions such as feeding, locomotion, or reproduction (Shubin et al. 1997). In insects, the presence of flexible joints makes possible the articulation of the otherwise rigid limbs. In fact, it is the presence of articulated appendages that give arthropods their name (from Greek *árthron*, "joint," and *pous*, "feet"). The *Drosophila* legs are composed of ten segments (from proximal to distal, *coxa*, *trochanter*, *femur*, *tibia*, five *tarsal* segments (*ta1 to ta5*), and *pretarsus*) each one separated by a flexible joint (Fig. 7.1a). The proximo-distal (P-D) axis of the

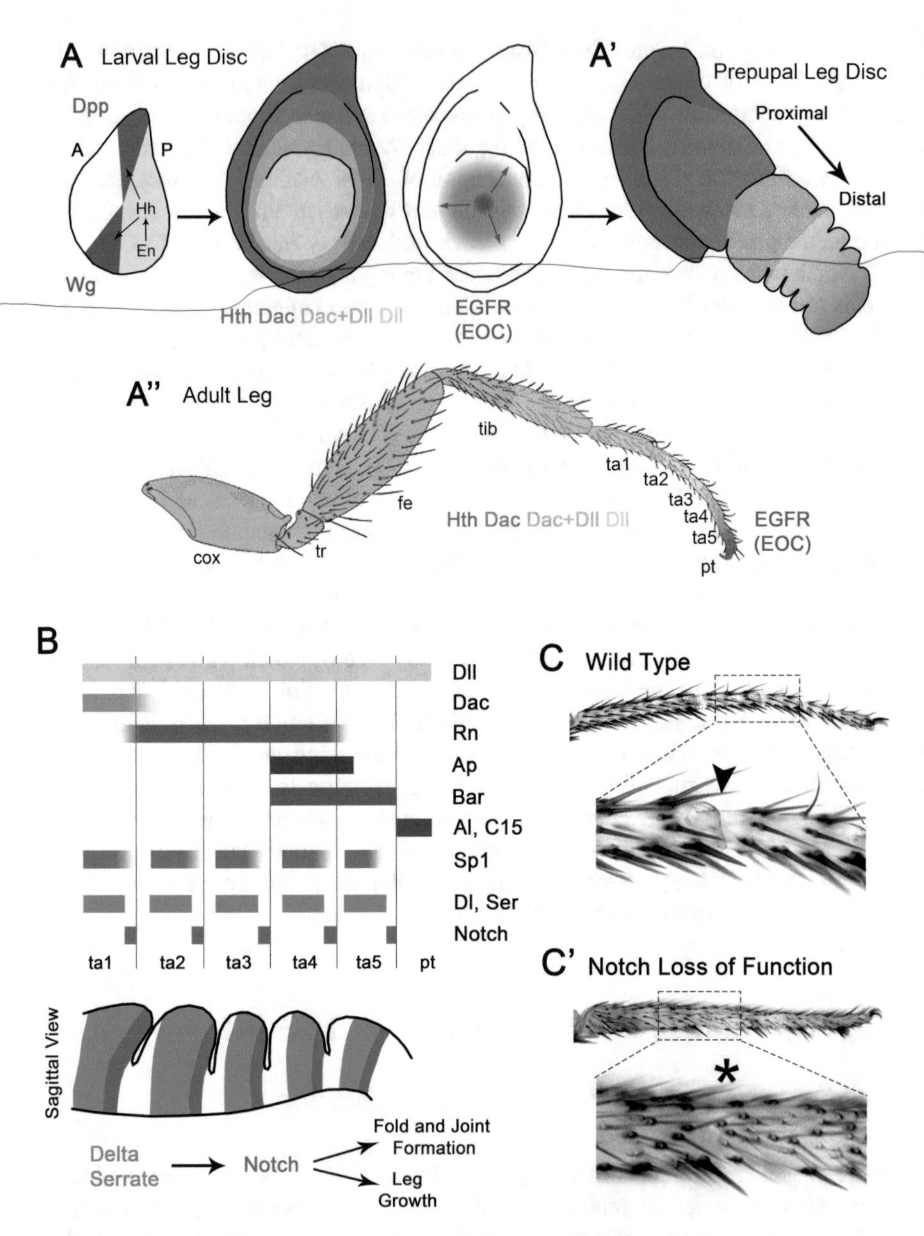

Fig. 7.1 P-D patterning of the leg directs Notch activity positioning. (**a**) Left: *Hh* is expressed in response to En in the posterior compartment and activates *dpp* (dorsal) and *wg* (ventral) expression in an early leg disc (left). Center: the expression of *Hth*, *dac*, and *Dll* in concentric rings set up the P-D patterning of the leg disc. A gradient of EGFR activity (EOC) in the center of the developing leg disc further refines distal patterning. (**a'**) Right: prepupal leg disc everts forming a tubelike structure, and the folds that position the future joints are already visible. (**a''**) Representation of the P-D patterning genes in an adult leg. (**b**) Tarsal region of a prepupal leg disc. The code generated by P-D genes serves as a blueprint for *Dl*/*Ser* expression and Notch activity positioning. Notch regulates joint formation and leg growth. (**c, c'**) Adult legs (tarsal region and close up of a joint below) of a *wild-type* (**c**) and a *Notch* hypomorph mutant (**c'**) fly

appendages is formed orthogonally to the anterior-posterior (A-P) and dorsal-ventral (D-V) axes that are already present in the embryo (Nusslein-Volhard and Wieschaus 1980; Reeves and Stathopoulos 2009; Estella et al. 2012).

Drosophila is a holometabolous insect that transits through four life stages (embryo, larva, pupa, and adult). Appendages arise from specific primordia, groups of ectodermal cells genetically specified during embryogenesis, that will invaginate to form the larval structures known as imaginal discs (Bate and Arias 1991; Cohen et al. 1993; McKay et al. 2009; Beira and Paro 2016). Imaginal discs are saclike epithelial monolayers that grow and become patterned during larval stages and will undergo metamorphosis throughout pupation to give rise to most adult cuticular structures, i.e., legs, wings, or eye and antenna (Beira and Paro 2016; Fristrom and Fristrom 1993; von Kalm et al. 1995; Pastor-Pareja et al. 2004; Aldaz et al. 2010; Ruiz-Losada et al. 2018).

The leg imaginal disc is divided into anterior and posterior compartments by the expression of the selector gene *engrailed* (*en*) in its posterior half, and this compartmental division is maintained throughout leg development. Posterior (*en*-positive) cells secrete the short-range ligand Hedgehog (Hh) that signals to adjacent anterior compartment boundary cells to activate the expression of two long-range signaling molecules, *decapentaplegic* (*dpp*) and *wingless* (*wg*). While *dpp* expression occurs in the dorsal half of the leg disc and determines dorsal identity, *wg* is restricted to the ventral half and confers ventral identity (Struhl and Basler 1993; Wilder and Perrimon 1995; Johnston and Schubiger 1996; Morimura et al. 1996; Theisen et al. 1996; Svendsen et al. 2015). Both Dpp and Wg inputs act combinatorially to establish the initial P-D axis of the leg discs. A cascade of transcription factors and the cross-regulation between them is responsible for the elaboration of the P-D axis of the leg. Briefly, high levels of Wg and Dpp activate *Dll* expression in the center of the leg disc. Dll activate another P-D patterning gene, *dachshund* (*dac*), which is repressed in the distal domain by peak levels of Wg and Dpp, allowing the formation of the Dll only domain. Dac then represses *Dll* expression in the medial region of the leg disc (Campbell et al. 1993; Diaz-Benjumea et al. 1994; Lecuit and Cohen 1997; Estella et al. 2008; Estella and Mann 2008; Giorgianni and Mann 2011). In the periphery of the leg discs, where low levels of combined Wg and Dpp are present, a third P-D patterning gene, *homothorax* (*hth*), is expressed (González-Crespo and Morata 1996; Abu-Shaar and Mann 1998; Gonzalez-Crespo et al. 1998). Therefore, the concentric expression domains from proximal to distal of *hth*, *dac*, *dac* + *Dll*, and *Dll*, also known as the "leg gap" genes, broadly define the P-D axis of the leg (reviewed in Estella et al. (2012), Ruiz-Losada et al. (2018)) (Fig. 7.1a).

Segmentation of the Tarsal Region

The distal-most region of the leg disc will give rise to the adult tarsal region and is further patterned during the last stage of larval development by a highly dynamic and complex genetic regulatory network (reviewed in Suzanne (2016), Kojima (2017)). Interestingly, while the number of proximal segments (coxa, trochanter, femur, and tibia) is conserved among arthropods, the number of tarsal segments is variable among species (Snodgrass 1935; Natori et al. 2012; Angelini et al. 2012). Nevertheless, the presence of five tarsal segments, as occurs in *Drosophila*, is thought to be the plesiomorphic state in insects (Grimaldi and Engel 2005). The regulatory cascade that governs tarsal region segmentation in *Drosophila* is highly dynamic (Natori et al. 2012), and simple changes in timing or extent of gene expression could potentially explain the variability of tarsal number in other insect species (Kojima 2017).

Patterning of the tarsal region is initiated by the activity of the epidermal growth factor receptor (EGFR) pathway at the center of early third instar leg discs (Campbell 2002; Galindo et al. 2002). High levels of Wg and Dpp activate the expression of the EGFR ligand Vein (Vn) and the protease rhomboid (Rho), which is necessary for ligand activation (Campbell 2002; Galindo et al. 2002, 2005; Newcomb et al. 2018). The activation of the EGFR pathway in a distal to proximal gradient creates the EGFR organizing center (EOC) that induces the nested expression in concentric domains of different tarsal patterning genes in a concentration-dependent manner (Fig. 7.1a). However, a recent study has shown that other sources of EGFR activity besides the EOC are also responsible to pattern the tarsal segments, as removing EOC activity from the distal tip only causes local P-D defects (Newcomb et al. 2018).

The end result of the EGFR activity and the cross-regulation between the different transcription factors during third instar leg imaginal disc development is the subdivision of the tarsal domain in five segments (ta1-ta5) and the pretarsus (Natori et al. 2012) (reviewed in Suzanne (2016), Kojima (2017)). For example, the future pretarsal region expresses the homeodomain transcription factors *aristaless* (*al*) and *C15*, while *Bar*, another homeodomain transcription factor, is expressed in the future ta5 region and weakly in the ta4 region. The ta4 region is defined by *apterous* (*ap*), a LIM-homeodomain transcription factor (Pueyo et al. 2000). Meanwhile, the tarsal-specific expression of *tarsal-less* (*tal*), *spineless* (*ss*), *rotund* (*rn*), and *bric-à-brac* (*bab*) further patterns the tarsal region (Natori et al. 2012; Godt et al. 1993; Duncan et al. 1998; Chu et al. 2002; Kozu et al. 2006; Pueyo and Couso 2008; Baanannou et al. 2013) and reviewed in Suzanne (2016), Kojima (2017) (Fig. 7.1b).

In summary, the final result of the interplay between "leg gap" genes, EGFR pathway activation, and tarsal-specific transcription factors is the subdivision of the leg disc in discrete regions of gene expression that give identity to the future adult segments of the leg (reviewed in Suzanne (2016), Kojima (2017)).

Dl/Ser Positioning and Notch Activation

Besides giving identity to the adult segments, P-D patterning of the leg disc acts as a positional blueprint for leg segmentation. The combined expression domains of "leg gap" genes and tarsal patterning genes generate a code of transcription factors that determines the positioning of the Notch ligands Dl and Ser in a band of cells at the distal end of each presumptive leg segment as the leg imaginal disc grows

(Rauskolb 2001) (Fig. 7.1b). How this positional information (i.e., a different combination of transcription factors in each segment) is integrated at the molecular level to activate *Dl* and *Ser* expression in each tarsal segment is largely unknown. However, the presence of dedicated *cis*-regulatory elements (CRMs) within *Ser* regulatory region that integrate positional information in each segment has been described (Rauskolb 2001; Cordoba et al. 2016). In this manner, it has been shown that Hth and Dac are required for the positioning of proximal rings of *Ser* expression, while the transcription factor Sp1 acts together with the tarsal-specific P-D transcription factor Ap to promote *Ser* expression in the *ta4* tarsal segment (Rauskolb 2001; Cordoba et al. 2016). It is therefore reasonable to think that the same logic could apply for the positioning of Ser and Dl in the remaining leg segments, despite the specific molecular details are yet to be elucidated.

The spatial localization of the Dl/Ser ligands elicits the activation of the Notch receptor in nine rings along the P-D axis. This is a key step in leg development, as Notch activity is required for the formation of the joints that separate each adult segment and for the correct growth of the appendage. Legs mutant for Notch or its ligands present fusions and reduction in the size of leg segments, whereas ectopic Notch activity induces the folding of the cuticle resembling joint formation (Angelini et al. 2012; de Celis et al. 1998; Bishop et al. 1999; Rauskolb and Irvine 1999; Cordoba and Estella 2014). Interestingly, each band of Dl/Ser-positive cells activates Notch signaling only in the adjacent cells located distally to them. The asymmetric distribution of the planar cell polarity (PCP) core proteins prevents Notch to be activated in the proximal side of Dl/Ser stripes. These PCP proteins are associated to the cell membrane and coordinate the orientation of the cells and cell structures along the plane of a tissue (Devenport 2014). In the leg imaginal epithelium, the cytoplasmic protein Dishevelled (Dsh) and the transmembrane protein Frizzled (Fz) are preferentially localized to the distal edge of each cell. Conversely, another transmembrane protein, Van Gogh (Vang), is located at the proximal side of the cells. Importantly, the direct interaction between Dsh and Notch contributes to the inhibition of Notch signaling (Axelrod et al. 1996; Munoz-Descalzo et al. 2010). Therefore, in the proximal adjacent cells to Dl/Ser, distally located Dsh blocks Notch signaling, whereas in those cells distal to Dl/Ser, Dsh is not present in their proximal side, allowing the interaction between the ligand and Notch receptor and promoting pathway activation (Capilla et al. 2012). Accordingly, flies mutant for the core PCP components result in a double band of Notch activity at both sides of Dl/Ser and adult tarsal legs with duplicated joints (Capilla et al. 2012).

The bidirectional signaling of the Notch pathway in the leg segments is also prevented by additional mechanisms. It has been described that another round of EGFR activity prevents the activation of Notch in the proximal cells adjacent to Dl/Ser in each segment (Galindo et al. 2005). Moreover, the expression of the transcription factor Defective proventriculus (Dve) in the inter-joint region is also required to prevent proximal activation of Notch (Shirai et al. 2007). This role could be the result of Dve-mediated repression of Notch target genes. Nevertheless, what is the relationship between Dve, EGFR, and the PCP pathway is mostly unknown.

The formation of sharp borders of gene expression is paramount for animal development (Dahmann et al. 2011). Indeed, given the relevance of Notch for leg development, there are multiple mechanisms that ensure precise localization of Notch activity, which often implies the function of transcription factors regulated in response to Notch that feed back to refine Notch pathway activation. The transcription factor *dAP-2* is a target of Notch that is expressed in all leg segments, and its loss causes the absence of joint structures and leg shortening (Monge et al. 2001; Kerber et al. 2001; Ahn et al. 2011). dAP-2 function seems to be primarily through the repression of Notch ligands Dl/Ser, in combination with Dve, to ensure precise borders of Notch signaling (Ciechanska et al. 2007).

Other mechanisms of Notch refinement are nevertheless restricted to certain regions of the leg. In the proximal leg, the interplay between the nuclear protein Lines and Brother of odd with entrails limited (Bowl), a member of the *odd-skipped* family of transcription factors, in response to Notch is confined to the so-called "true" joints (see below). *drumstick* (*drm*), another *odd-skipped* family member, is expressed in Notch-positive cells and inhibits Lines-mediated degradation of Bowl. Bowl, then, represses *Dl* expression in the Notch activation domain, thus defining a sharp border of Notch activity (Greenberg and Hatini 2009). A similar mechanism, but governed by the short peptides encoded by the noncanonical gene *tal*, has been proposed for the regulation of Dl/Notch boundaries in the tarsal region of the leg. In this case, *tal* is expressed in Notch-positive cells and represses *Dl* through the activity of the Shavenbaby (Svb) transcription factor (Pueyo and Couso 2011). Also at the tarsal region, the *zinc finger homeodomain-2* (*zfh-2*) gene is implicated in maintaining Notch activity, possibly through the regulation of *tal* (Guarner et al. 2014).

In summary, the P-D segmental identity of the leg provided by a complex interplay of transcription factors and regulatory networks is eventually translated in the precise positioning of Notch ligands Dl and Ser. The ensuing activation of Notch pathway, which is asymmetrically restricted by PCP and refined by multiple feedback mechanisms, is then instructive for the formation of the joints and for the correct growth of the leg disc. In the next section, we will summarize the molecular and cellular mechanisms acting downstream of Notch that direct joint morphogenesis and growth control.

Notch Regulation of Leg Development

Joint Formation in Response to Notch Activity

At the end of larval development, the stripes of Notch-positive cells that will determine the exact position of joint development are already set along the P-D axis, in the distal-most region of each presumptive leg segment. However, the shape changes associated with joint morphogenesis will not be evident until the onset of pupal development, while the characteristic adult joint structure will form in the late pupa. How the morphogenetic mechanisms that shape the joints are orchestrated in response to Notch activity is the object of this section.

Although all adult joints appear similar at first glance, they are not all equivalent in terms of their morphology, their evolutionary origin, or the developmental mechanisms that shape them (Snodgrass 1935). Attending to morphologic criteria, leg joints can be broadly subdivided into two classes. "True" or proximal joints are characterized by an asymmetrical architecture and the presence of attached muscles and tendons, which makes them motile. This type of joints is found in the proximal segments of the leg and in the tarsus-pretarsus interface (Snodgrass 1935; Mirth and Akam 2002; Soler et al. 2004) (Fig. 7.2a–c). "Tarsal" or distal joints present a ball-and-socket structure that is devoid of tendon or muscle attachments and are found within the tarsal region (Mirth and Akam 2002; Tajiri et al. 2010, 2011). Interestingly, it has been proposed that the ground state of a ventral appendage in the absence of leg and antenna selector gene activity consists of a unique proximal segment and a tarsal region containing five tarsi (Casares and Mann 2001). This data and the variability in the number of segments of the tarsal region, which is not observed within the proximal segments, point to different evolutionary origins for the proximal and tarsal regions of the leg (Kojima 2017; Natori et al. 2012; Angelini et al. 2012).

Thus, in order to give rise to these different joint typologies, Notch function bifurcates into, at least, two different downstream genetic regulatory programs to form proximal and tarsal joints. Accordingly, the restricted expression and activity of Notch target genes in either proximal or tarsal joints have been reported. It is therefore possible that the function of such genes eventually regulates the different developmental programs that shape both types of joint.

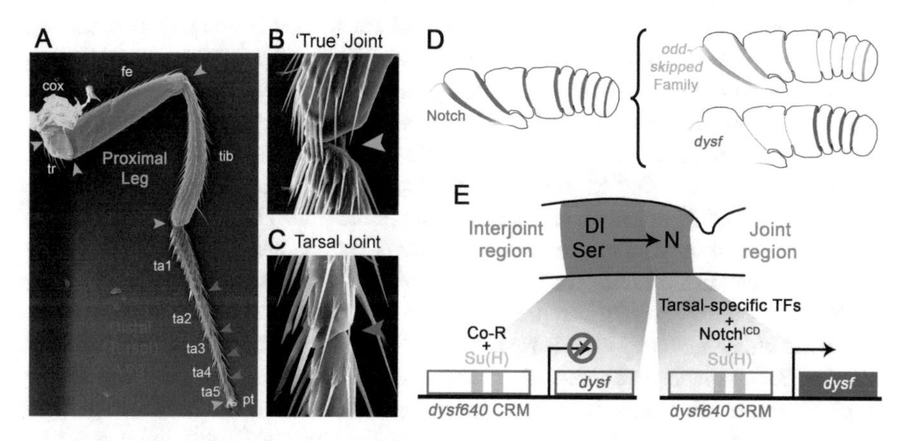

Fig. 7.2 Development of proximal and tarsal joints. (**a**) Scanning electron image of an adult leg, showing the proximal and distal regions of the leg. (**b** and **c**) Detail of a proximal or "true" joint and distal or "tarsal" joint (arrowheads). (**d**) Notch is activated in each presumptive leg joint in a prepupal leg disc. The *odd-skipped* genes are expressed exclusively at the presumptive proximal joints, while *dysf* is expressed only in the presumptive tarsal joints. (**e**) *dysf* expression is regulated by Notch through the direct binding of Su(H) to two dedicated binding sites at *dysf640* CRM. In the absence of Notch (inter-joint region), Su(H) associates with corepressors (Co-R) to inhibit *dysf* expression. When Notch is active (joint region), Co-R are displaced by $Notch^{ICD}$ allowing Su(H) to activate *dysf* transcription. Tarsal-specific transcription factors restrict *dysf* expression to the distal leg

The members of the *odd-skipped* family of transcription factors *odd-skipped* (*odd*), *sister of odd and bowl* (*sob*), and *drumstick* (*drm*) are specifically expressed at the presumptive proximal joints in response to Notch (Hao et al. 2003) (Fig. 7.2d). Ectopic expression of *odd*, *sob*, or *drm* in the leg imaginal disc is sufficient to form ectopic folds in the epithelium, as well as cuticle indentations in the adult leg that resemble joint formation. Nevertheless, the elimination of *odd*, *sob*, or *drm* function individually does not inhibit joint formation, suggesting that these genes act redundantly during leg development (Hao et al. 2003). Accordingly, the combined downregulation of *odd, sob*, and *drm* in the beetle *Tribolium castaneum* results in the loss of proximal joints (Angelini et al. 2012). Nevertheless, the molecular and cellular mechanisms that are regulated by the *odd-skipped* family members to prompt epithelial folding and joint formation are largely unknown. Interestingly, another *odd-skipped* family member, *bowl*, has a wider expression pattern, and its loss of function suggests that its function is not restricted to proximal joints, but is rather needed for specification and segmentation of the tarsal region (Hao et al. 2003; de Celis Ibeas and Bray 2003). The POU-domain transcription factor Nubbin (Nub) is also expressed in response to Notch activity and restricted to the true joint domain (Ng et al. 1995; Rauskolb et al. 1999). It has been reported that strong mutant *nub* alleles cause the development of shorter and gnarled legs (Cifuentes and Garcia-Bellido 1997). *nub* expression is not altered in *odd* gain-of-function experiments, indicating that its role on leg formation downstream of Notch is independent from the function of the *odd-skipped* family members (Hao et al. 2003).

Conversely, there are genes which expression pattern is restricted to the presumptive tarsal joints and that might act downstream of Notch to regulate tarsal-specific joint morphogenesis. One of the best-characterized Notch targets in the leg is the bHLH-PAS transcription factor Dysfusion (Dysf) (Fig. 7.2d). *dysf* expression is restricted to the presumptive four tarsal joints where it overlaps with the known Notch target genes *big brain* (*bib*) and *Enhancer of split mβ* (*E(spl)mβ*) (de Celis et al. 1998). A specific *cis*-regulatory module for *dysf* (named *dysf640* CRM) that faithfully reproduces *dysf* expression in the leg has been identified. Molecular dissection of this CRM revealed, at least, two functional Su(H) binding sites that allow direct regulation in response to Notch activity (Cordoba and Estella 2014). In the absence of Notch signaling (i.e., at the inter-joint domains of the leg segments), Su(H) form a complex with corepressors, keeping *dysf* expression silent (Fig. 7.2e). When the Notch pathway is activated, Notch^{ICD} translocates to the nucleus, where it binds to Su(H) displacing corepressors and activating *dysf* expression at high levels in presumptive joint cells. Importantly, for *dysf* expression to be restricted only to the tarsal domain, an additional layer of regulation besides Notch input must be present to dictate precise P-D localization. This regulation is made evident upon mutation of the Su(H) binding sites in the *dysf640* CRM reporter construct. In this case, Notch is unable to activate *dysf* expression at the presumptive joint cells, and the repression exerted by Su(H) at the inter-joint domain is also lost. Mutation of Su(H) binding sites in the *dysf640* CRM leads to uniform low levels of reporter gene expression throughout the tarsal region (Cordoba and Estella 2014). Nevertheless, the specific P-D gene or genes that provide this tarsal restriction to *dysf* expression are yet to be described. Therefore, the *dysf* CRM is a logic integrator of Notch signaling and a P-D input that restricts *dysf* expression to the presumptive tarsal joints. This mode of transcriptional regulation by Notch is compatible with the *default repression* model widely employed by numerous signaling pathways and specifically by Notch (Barolo et al. 2002; Lai 2002). This model proposes that the transcriptional effectors of signaling pathways activate transcription upon pathway activation in a context-specific manner while repress transcription in the absence of signaling (Affolter et al. 2008).

Consistently with *dysf* expression pattern in the tarsal leg disc, adult flies mutant for *dysf* completely lose all tarsal joints, whereas proximal joints remain unaffected. Despite *dysf* mutants display joint defects that resemble Notch loss-of-function phenotypes, the Notch pathway is correctly localized and activated throughout the tarsal domain in these mutants. Moreover, *dysf* misexpression causes cuticular folding that resembles ectopic joint-like structures in the adult leg and is capable of doing so even in the absence of Notch activity (Cordoba and Estella 2014). Therefore, *dysf* is a bona fide Notch target gene during leg tarsal joint development, which is completely necessary for tarsal joint development.

Several genes besides *dysf* are also exclusively expressed in the presumptive tarsal joints such as *tal*, *deadpan* (*dpn*), or *Pox neuro* (*Poxn*). *tal* is expressed in response to Notch and ensures sharp Notch borders by repressing *Dl* expression (Pueyo and Couso 2011). *dpn*, as *dysf*, also encodes for a bHLH transcription factor expressed in the four tarsal joints, although no Dpn role has been reported in the leg yet. *Poxn* is required for the formation of a subset of tarsal joints; however its relationship with Notch or *dysf* needs to be studied (Awasaki and Kimura 2001).

Dysfusion Control of Tarsal Joint Morphogenesis

As we have described before, the Notch downstream regulators that differentially direct joint formation in tarsal and proximal joints have been identified. Nevertheless, the study of the molecular and cellular processes that execute the morphogenetic changes required to form a joint has been mainly performed in the tarsal region (Suzanne 2016; Kojima 2017; Tajiri et al. 2010), whereas much less is known about the formation of proximal joints (Fristrom and Fristrom 1993; Mirth and Akam 2002).

The process of tarsal joint morphogenesis can be roughly divided in three phases: an initial sharp epithelial folding, a partial unfolding, and a posterior re-constriction of the epithelium to form the characteristic adult joint structure. The first physical evidence of tarsal joint formation is observed during prepupal development (0–6 h after puparium formation), as the imaginal leg discs telescope out to form a cylinder where the central region of the disc becomes the distal-most portion of the leg (see Fig. 7.1). At this stage, four deep epithelial folds, transversal to the P-D axis of the leg, are formed by bands of cells that undergo apical constriction. These constrictions form just distally to Notch-activating cells in each presumptive tarsal segment (de Celis et al. 1998; Cordoba and Estella 2014; Manjon et al. 2007; Monier et al. 2015) (Fig. 7.3). Later during pupal development, these folds are partially flattened while the leg elongates, although certain degree of apical constriction is still maintained

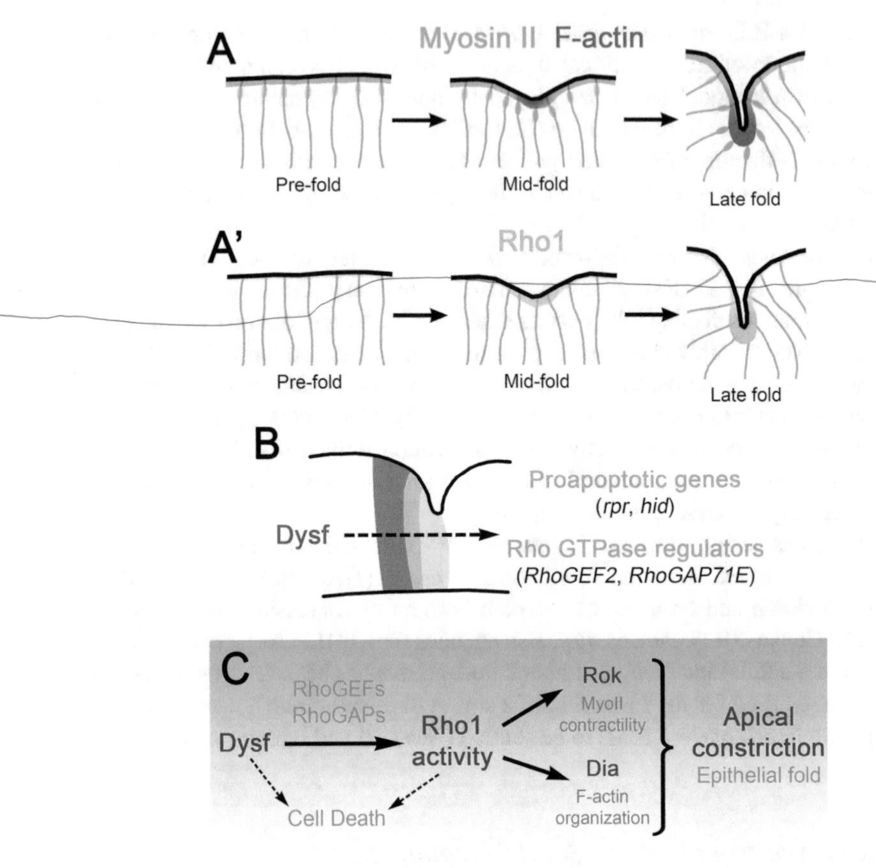

Fig. 7.3 Molecular and cellular mechanisms of tarsal fold formation. (**a**, **a'**) Schematic representation of F-actin and Myosin II (**a**) and Rho1 (**a'**) localization as apical constriction and tarsal fold formation progresses. (**b**) Dysf regulates the transcription of proapoptotic genes and Rho GTPase regulators. Note that Dysf domain and its transcriptional targets are not completely coincident. (**c**) Proposed model for epithelial fold formation in response to Dysf activity

(Mirth and Akam 2002). Afterward, starting at 24 h after puparium formation, another round of morphogenesis that again requires Notch activity forms the final ball-and-socket structure of the adult joint (Kojima 2017; Mirth and Akam 2002; Tajiri et al. 2010, 2011). In the next sections, we will focus on the initial steps of tarsal joint morphogenesis, which occur during prepupal development.

Localized Apical Constriction Drives Epithelial Folding to Form the Tarsal Joints

Apical constriction is a conserved and widespread mechanism that is reiteratively used in animal development for epithelial tissue morphogenesis. It consists on the shrinkage of the apical surface of either individual cells, usually leading to delamination (An et al. 2017), or of groups of cells leading to tissue folding or invagination (Sawyer et al. 2010; Martin and Goldstein 2014). In *Drosophila,* several developmental processes that use apical constriction have been thoroughly studied, and specifically embryonic gastrulation has been greatly informative to unveil the molecular mechanisms that cause these cell shape changes (Leptin and Grunewald 1990; Martin et al. 2009; Weng and Wieschaus 2016; Gilmour et al. 2017).

Apical constriction relies in the tight coordination between the cytoskeleton activity and the cell adhesion components that generate and transmit the force to neighbor cells, respectively. The contraction of actin filament (F-actin) networks by the non-muscle Myosin II (Myo II) generates the force that drives apical constriction (Martin et al. 2009; Roh-Johnson et al. 2012). Myo II is a hexameric protein composed by two regulatory light chains (encoded by the gene *spaghetti squash* (*sqh*)), two heavy chains (encoded by *zipper* (*zip*)), and two essential light chains (encoded by *Myosin light chain cytoplasmic* (*Mlc-c*)). Myo II is activated through phosphorylation of Sqh, and its motor activity resides on the function of Zip, which pulls on F-actin to generate contractile force (Karess et al. 1991; Tan et al. 1992). This force has to be efficiently transmitted to the neighboring cells through the binding of F-actin at the level of cell-cell adhesion domains (Martin et al. 2010; Mason et al. 2013; Marston et al. 2016; Vasquez and Martin 2016). An important player in coordinating cytoskeleton dynamics and cell adhesion in *Drosophila* is the formin protein Diaphanous (Dia), which facilitates the assembly of F-actin at the level of adherens junctions (Mason et al. 2013; Homem and Peifer 2008; Liu et al. 2010).

During epithelial folding at the presumptive tarsal joints of the leg imaginal disc, bands of cells coordinately undergo apical constriction, reducing their apical surface while accumulating high levels of F-actin at their apical domains (Monier et al. 2015; Cordoba and Estella 2018). Importantly, in *dysf* loss-of-function conditions, this F-actin accumulation is lost, together with the formation of epithelial folds, suggesting a functional relationship between both processes (Cordoba and Estella 2018). Meanwhile, levels of the motor protein Myo II are also incremented in the apical region of fold-forming cells, where it is accumulated at the level of the adherens junctions (Monier et al. 2015; Cordoba and Estella 2018). It would be interesting to study in detail the organization and dynamics of actomyosin networks during tarsal epithelial joint formation and compare its molecular mechanisms with better-known models of apical constriction such as ventral furrow formation (Martin et al. 2009; Mason et al. 2013; Dawes-Hoang et al. 2005) (Fig. 7.3).

Dysf-Dependent Rho1 Activity Regulates Apical Constriction

At the regulatory level, apical constriction is globally controlled by the activity of the Rho GTPase Rho1 (Martin and Goldstein 2014). The Rho family of small GTPases functions as molecular switches controlling a wide number of cellular functions including cell division, cell adhesion, apicobasal polarity, cell shape changes, or cell migration. Rho GTPases play a central role in the reorganization of the actin cytoskeleton and therefore their activity is tightly regulated (Jaffe and Hall 2005; Bausek and Zeidler 2014; Citi et al. 2014; Mack and Georgiou 2014; Zegers and Friedl 2014). The most prominent Rho GTPases are Rho (Rho1 in *Drosophila*), Rac, and Cdc42, which play conserved roles between *Drosophila* and vertebrates.

In particular, Rho1 has been reported to control both the generation of contractile force and the control of F-actin assembly at adherens junctions (Martin and Goldstein 2014; Gilmour et al. 2017; Jaffe and Hall 2005; Zegers and Friedl 2014). A well-established target of Rho1 activity is the Rho-associated protein kinase (Rok) that activates Myo II contractility through phosphorylation of Sqh (Winter et al. 2001; Boettner and Van Aelst 2002; Riento and Ridley 2003; Xu et al. 2008). In *Drosophila* another kinase, Death-associated protein kinase related (Drak), also phosphorylates Sqh, and its function becomes necessary when Rok activity is compromised (Neubueser and Hipfner 2010; Robertson et al. 2012). Additionally, the formin Dia, another downstream target of Rho1, links actin cytoskeleton to adherens junctions providing a molecular framework for Rho1 global control of morphogenesis, including apical constriction (Mason et al. 2013; Homem and Peifer 2008; Mulinari et al. 2008; Kuhn and Geyer 2014).

During tarsal joint development, Rho1 protein appears specifically localized around the apical region of the cells that form the epithelial tarsal folds (Cordoba and Estella 2018) (Fig. 7.3). Using a reporter construct for Rho1 activity (Simoes et al. 2006), it has been shown that this localization is coupled with increased Rho1 activity around these folds (Cordoba and Estella 2018). Importantly, both patterned Rho1 localization and activity around the forming folds are lost in *dysf* knockdown imaginal discs, indicating that Dysf ultimately regulates Rho1 positioning and activity levels during tarsal fold formation (Cordoba and Estella 2018). Moreover, direct blocking of Rho1 activity inhibits epithelial fold and adult joint formation, phenotypes similar to *dysf* loss of function (Cordoba and Estella 2014, 2018). As expected, knockdown of the Rho1 effectors Rok, Drak, and Dia also disrupts epithelial folding and adult joint formation to different degrees, probably due to the functional redundancy between them (Cordoba and Estella 2018; Neubueser and Hipfner 2010; Robertson et al. 2012). Therefore, Dysf control of tarsal fold morphogenesis is executed through its regulation of precise spatial activation of Rho1 that, in turn, coordinates apical constriction in the leg imaginal disc.

Dysf Transcriptional Control of Apical Constriction

We have described in this review that Dysf controls the cellular mechanisms implicated in apical constriction during tarsal epithelial fold formation, specifically Rho1 localization and activity. Nevertheless, as a transcription factor, Dysf should control these cellular mechanisms at the level of transcriptional regulation, through the activation of a specific set of effector target genes. Interestingly, Dysf has been reported to transcriptionally regulate both Rho GTPase regulators and proapoptotic genes, effectors that could promote Rho1 activity and apical constriction ((Cordoba and Estella 2014; Manjon et al. 2007; Monier et al. 2015; Greenberg and Hatini 2011) and reviewed in Suzanne (2016)).

Rho1 GTPase functions as a molecular switch that cycles between active (GTP-bound) and inactive (GDP-bound) conformational states. These transitions are modulated by the activity of guanine exchange factors (GEFs) and GTPase-activating proteins (GAPs) that promote the activation and inactivation, respectively, of Rho GTPases (Jaffe and Hall 2005). The precise regulation of Rho1 activity in a cell is mediated by an exquisite balance between RhoGEF and RhoGAP protein levels in the cytoplasm (Simoes et al. 2006). Interestingly, Rho GTPase regulators have been previously implicated in the control of apical constriction and epithelial folding (Mulinari et al. 2008; Simoes et al. 2006; Fox and Peifer 2007; Kolsch et al. 2007; Mason et al. 2016). In a comprehensive analysis of Rho GTPase regulators expressed in the *Drosophila* leg imaginal disc, Greenberg and Haitini described at least 17 *RhoGEFs* and *RhoGAPs* expressed in the presumptive leg joints (Greenberg and Hatini 2011). A subset of them, which includes *RhoGEF2*, *RhoGAP71E*, and *RhoGAP68F*, are restricted to the tarsal region, and their expression is regulated downstream of Notch signaling (Greenberg and Hatini 2011). Interestingly, knocking down the activity of some of these RhoGEFs and RhoGAPs causes defects in leg disc epithelial folding and joint morphogenesis (Greenberg and Hatini 2011; de Madrid et al. 2015). A detailed comparison between *RhoGEF2*, *RhoGAP71E*, and *dysf* expression at the tarsal presumptive joints of the leg imaginal disc indicates that their localization partially overlaps. Expression of *RhoGEF2* and *RhoGAP71E* extends a couple of cell rows distally of the Dysf domain toward the fold (Fig. 7.3). Despite the non-complete coincidence between Dysf and RhoGEF2/RhoGAP71E localization, Dysf is absolutely necessary for their expression (Cordoba and Estella 2014).

Apoptosis or programmed cell death has been proposed to play an important morphogenetic role sculpting structures such as vertebrate digits or fly embryonic segments ((Lohmann et al. 2002; Suzanne et al. 2010; Hernandez-Martinez and Covarrubias 2011; Yamaguchi et al. 2011) and reviewed in Suzanne and Steller (2013)). In the *Drosophila* leg imaginal disc, localized expression of the proapoptotic genes *reaper* (*rpr*) and *heads involution defective* (*hid*) and patterned apoptosis is observed around the presumptive tarsal joints (Cordoba and Estella 2014; Manjon et al. 2007; Monier et al. 2015). The expression of the proapoptotic genes relies on the c-Jun N-terminal kinase pathway (JNK) that is activated by a sharp Dpp border at the end of each tarsal segment (Manjon et al. 2007). Importantly, although *rpr* and *hid* expression domains are only partially overlapping with Dysf, just as occurs with *RhoGEF2* and *RhoGAP71E*, Dysf is necessary and sufficient for their expression in the tarsal region (Cordoba and Estella 2014) (Fig. 7.3).

A functional role for apoptosis in the *Drosophila* leg was proposed after the analysis of defective epithelial folding and joint formation in several conditions of cell death inhibition ((Manjon et al. 2007; Monier et al. 2015) and reviewed in Suzanne (2016)). According to the mechanistic model proposed for apoptosis-driven epithelial folding, dying cells generate an apicobasal force that causes transient indentations in the epithelium and Myo II accumulation in the neighboring cells (Monier et al. 2015; Ambrosini et al. 2019). The temporal and spatial coordination of individual apoptotic events in bands around the leg disc circumference would cause the epithelium to fold in a stable manner (Monier et al. 2015). Interestingly, apoptosis involves extensive cytoskeletal remodeling, and the activation of Rho-dependent actomyosin contractility in its neighbors is necessary for correct cell extrusion ((Rosenblatt et al. 2001; Slattum et al. 2009) and reviewed in Monier and Suzanne (2015)). Therefore, it could be proposed that the transcriptional regulation of the proapoptotic genes by Dysf leads to Rho1 GTPase activation that induces the folding of the epithelium through Myo II activity.

Nevertheless, a more recent study failed to observe defects in fold or adult joint formation when apoptosis was eliminated from the entire tarsal domain. In these apoptotic-deficient conditions, apical constriction, apical F-actin accumulation, and Rho1 activity at the epithelial folds remained unaffected (Cordoba and Estella 2018). These results strongly suggest that apoptosis does not play an instructive role in tarsal epithelial fold and joint development. Thus, it is possible that the observed localized apoptosis is a consequence of tissue remodeling at the tarsal folds, rather than the driving force behind folding (Fig. 7.3). Interestingly, Rho1 activity has been shown to induce apoptosis (Vidal et al. 2006; Neisch et al. 2010), suggesting that the localized cell death present at the tarsal region could appear as a result of increased Rho1 activity in epithelial fold cells. However, it cannot be ruled out a possible function of cell death in defining the detailed architecture of the joint or enhancing the dynamics of fold formation.

In conclusion, Notch controls leg morphogenesis through the differential activation of subsidiary transcription factors in the proximal and distal leg territories. In the tarsal region, Notch directly activates *dysf* expression that, through the modulation of Rho1 activity, promotes fold and joint formation. Dysf regulation of Rho1 activity provides a simple model to explain how cellular mechanisms that elicit epithelial fold morphogenesis are coordinated downstream of a transcription factor. Nevertheless, a detailed molecular explanation of the transcriptional control of Rho1 regulators by Dysf is yet to be provided.

Adult Joint Formation Depends on Notch Activity

As mentioned previously in this review, the differentiation and final shape acquisition of the adult joints occurs during the last stages of pupal development. The initial constrictions that prefigure the joints are almost completely unfolded after the prepupal stage as the leg disc elongates and reduce its diameter, although cells at the presumptive joints remain slightly constricted and are aligned, making them morphologically distinguishable (Fristrom and Fristrom 1993; Mirth and Akam 2002; Tajiri et al. 2010). The morphogenetic processes that give leg joints their definitive configuration start around 24 h after pupation for both proximal and tarsal joints (Fristrom and Fristrom 1993; Mirth and Akam 2002; Tajiri et al. 2010). At this stage, the joint epithelium undergoes shape changes that result in the cells at the proximal border of the joint to form a “lip” on top of distal cells. The initial stages of this process are similar between proximal and tarsal joints, while the characteristic asymmetries of proximal joints (i.e., the tibial-tarsal joint) arise later, by 36 h of pupal development (Mirth and Akam 2002). Interestingly, apical constriction of joint cells and apical F-actin accumulation is observed for both proximal and distal joints, suggesting that a mechanism similar to prepupal fold formation could be taking place (Fristrom and Fristrom 1993; Mirth and Akam 2002; Tajiri et al. 2010).

After this secondary constriction at the joints is set, the adult cuticle starts being deposited in the apical side of epithelial cells, at around 45 h after pupation. Nevertheless, the details of joint formation at this stage have only been studied to date in tarsal joints, which present a characteristic ball-and-socket morphology (Tajiri et al. 2010). The ball and the socket structures are located at opposite sides of the joint and present a defined and complementary shape that allows its smooth flexion. Distinct cell populations will form the ball and the socket in a sequential morphogenetic process that requires precise cell shape changes and the secretion of chitin to the extracellular matrix (ECM) in a highly stereotyped manner (Tajiri et al. 2010). Interestingly, continued Notch signaling from approximately 24–78 h of pupal development is necessary for the correct differentiation of the ball and socket (Tajiri et al. 2011). Notch function is necessary for both the fate specification of ball vs. socket cells and for the extensive cell shape changes that form these structures (Tajiri et al. 2011). Further investigation would be required to test whether Notch is also necessary for proximal joint development at these stages.

Notch Regulation of Leg Growth

Notch activity is required for leg segmentation and for the development of leg joints. Additionally, Notch also controls leg growth in a nonautonomous manner. When mutant clones for *Notch* or its ligand Dl that span two leg segments were generated, a clear reduction in leg size was observed that affected both mutant and adjacent wild-type tissues (de Celis et al. 1998; Rauskolb and Irvine 1999). Interestingly, the ectopic activation of the Notch pathway not only induces the formation of joint-like structures but also promotes outgrowths composed of wild-type and mutant cells (Rauskolb and Irvine 1999). Therefore, it is likely that Notch activates a downstream effector pathway that stimulates leg growth on adjacent leg segments.

One candidate to mediate this effect is the Hippo tumor suppressor pathway that controls organ size in insects and mammals (Halder and Johnson 2011). This pathway is controlled upstream by two atypical cadherin molecules, Dachsous and Fat (Ds-Fat), that form heterodimers, and by the Golgi-localized protein kinase four-jointed (Fj). Fat and Ds bind each other as receptor and ligand, whereas Fj modulates the interaction between them. In the fly, *ds* and *fj* viable mutants present reduced growth in the proximo-distal axis resulting in shorter legs (Villano and Katz 1995; Clark et al. 1995; Mao et al. 2006). Both *fj* and *ds* are expressed as complementary gradients in the imaginal discs and these gradients are essential for the correct growth and shape of these organs (Halder and Johnson 2011). The current model proposes that different levels of Ds/Fj between cells regulate the Hippo pathway and therefore the size of an organ (Halder and Johnson 2011). In the leg disc, the expression of *ds* and *fj* is more complex and dynamic, possibly reflecting the variety of sizes within the different leg segments. Although very little is known about how *ds*, *fat*, and *fj* regulate growth along the P-D axis of the *Drosophila* leg, in other arthropods such as the cricket Gryllus, *fj* and *ds* are expressed as proximal and distal gradients, respectively, in each leg segment (Bando et al. 2009). It is possible that the positional information generated by the P-D axis genes together with the activation of the Notch pathway at each presumptive leg joint could define the Ds/Fj gradients leading to a cell-to-cell signaling mechanism that regulates the growth and shape of the leg. Which are the molecular mechanisms that connect the Notch pathway with the Hippo pathway to regulate leg growth is mostly unknown; however several studies have pointed out this relationship. In the leg, *fj* is activated in *Dl* and *Ser* expressing cells and it is negatively regulated by Notch signaling (Rauskolb and Irvine 1999). In addition, Fj and Notch act in a feedback loop to refine their domain of activity (Buckles et al. 2001). Importantly, ectopic expression of *fj* induces the formation of outgrowths in the leg that are entirely composed of wild-type tissue, highlighting its role as a nonautonomous regulator of tissue growth (Buckles et al. 2001).

To summarize, the Notch pathway is activated at the boundaries between leg segments during *Drosophila* leg development, and it is required at different stages to direct joint formation and to control the correct growth of the leg. The precise spatial localization of the Notch pathway in the leg depends on the integration at the CRMs of *Dl/Ser* of the positional information provided by the P-D patterning genes. Notch directs the formation of the different leg joints by the spatially localized activation of subsidiary target genes. In the tarsal region, the Notch target Dysf controls actomyosin cytoskeleton dynamics and cell shape changes through the regulation of Rho1 during joint morphogenesis. Finally, although the molecular mechanisms that control leg growth by Notch are mostly unknown, some reports link Notch to the Hippo pathway to execute this function.

References

Abu-Shaar M, Mann RS (1998) Generation of multiple antagonistic domains along the proximodistal axis during Drosophila leg development. Development 125:3821–3830

Affolter M, Pyrowolakis G, Weiss A, Basler K (2008) Signal-induced repression: the exception or the rule in developmental signaling? Dev Cell 15:11–22. https://doi.org/10.1016/j.devcel.2008.06.006

Ahn Y, Zou J, Mitchell PJ (2011) Segment-specific regulation of the Drosophila AP-2 gene during leg and antennal development. Dev Biol 355:336–348. https://doi.org/10.1016/j.ydbio.2011.04.032

Aldaz S, Escudero LM, Freeman M (2010) Live imaging of Drosophila imaginal disc development. Proc Natl Acad Sci U S A 107:14217–14222. https://doi.org/10.1073/pnas.1008623107

Ambrosini A, Rayer M, Monier B, Suzanne M (2019) Mechanical function of the nucleus in force generation during epithelial morphogenesis. Dev Cell 50:197. https://doi.org/10.1016/j.devcel.2019.05.027

An Y et al (2017) Apical constriction is driven by a pulsatile apical myosin network in delaminating Drosophila neuroblasts. Development 144:2153–2164. https://doi.org/10.1242/dev.150763

Andersson ER, Sandberg R, Lendahl U (2011) Notch signaling: simplicity in design, versatility in function. Development 138:3593–3612. https://doi.org/10.1242/dev.063610

Angelini DR, Smith FW, Jockusch EL (2012) Extent with modification: leg patterning in the beetle tribolium castaneum and the evolution of serial homologs. G3 (Bethesda) 2:235–248. https://doi.org/10.1534/g3.111.001537

Awasaki T, Kimura K (2001) Multiple function of poxn gene in larval PNS development and in adult appendage formation of Drosophila. Dev Genes Evol 211:20–29

Axelrod JD, Matsuno K, Artavanis-Tsakonas S, Perrimon N (1996) Interaction between Wingless and Notch signaling pathways mediated by dishevelled. Science 271:1826–1832. https://doi.org/10.1126/science.271.5257.1826

Baanannou A et al (2013) Drosophila distal-less and Rotund bind a single enhancer ensuring reliable and robust bric-a-brac2 expression in distinct limb morphogenetic fields. PLoS Genet 9:e1003581. https://doi.org/10.1371/journal.pgen.1003581

Bando T et al (2009) Regulation of leg size and shape by the Dachsous/Fat signalling pathway during regeneration. Development 136:2235–2245. https://doi.org/10.1242/dev.035204

Barolo S, Stone T, Bang AG, Posakony JW (2002) Default repression and Notch signaling: hairless acts as an adaptor to recruit the corepressors Groucho and dCtBP to Suppressor of Hairless. Genes Dev 16:1964–1976. https://doi.org/10.1101/gad.987402

Bate M, Arias AM (1991) The embryonic origin of imaginal discs in Drosophila. Development 112:755–761

Bausek N, Zeidler MP (2014) Galpha73B is a downstream effector of JAK/STAT signalling and a regulator of Rho1 in Drosophila haematopoiesis. J Cell Sci 127:101–110. https://doi.org/10.1242/jcs.132852

Beira JV, Paro R (2016) The legacy of Drosophila imaginal discs. Chromosoma 125:573–592. https://doi.org/10.1007/s00412-016-0595-4

Bigas A, Espinosa L (2018) The multiple usages of Notch signaling in development, cell differentiation and cancer. Curr Opin Cell Biol 55:1–7. https://doi.org/10.1016/j.ceb.2018.06.010

Bishop SA, Klein T, Arias AM, Couso JP (1999) Composite signalling from Serrate and Delta establishes leg segments in Drosophila through Notch. Development 126:2993–3003

Boettner B, Van Aelst L (2002) The role of Rho GTPases in disease development. Gene 286:155–174

Bray S (1998) Notch signalling in Drosophila: three ways to use a pathway. Semin Cell Dev Biol 9:591–597. https://doi.org/10.1006/scdb.1998.0262

Bray SJ (2006) Notch signalling: a simple pathway becomes complex. Nat Rev Mol Cell Biol 7:678–689. https://doi.org/10.1038/nrm2009

Bray S, Bernard F (2010) Chapter eight – Notch targets and their regulation, Elsevier Inc. Curr Top Dev Biol. 92:253–275. https://doi.org/10.1016/S0070-2153(10)92008-5

Bray S, Furriols M (2001) Notch pathway: making sense of suppressor of hairless. Curr Biol 11:R217–R221

Buckles GR, Rauskolb C, Villano JL, Katz FN (2001) Four-jointed interacts with dachs, abelson and enabled and feeds back onto the Notch pathway to affect growth and segmentation in the Drosophila leg. Development 128:3533–3542

Campbell G (2002) Distalization of the Drosophila leg by graded EGF-receptor activity. Nature 418:781–785

Campbell G, Weaver T, Tomlinson A (1993) Axis specification in the developing Drosophila appendage: the role of wingless, decapentaplegic, and the homeobox gene aristaless. Cell 74:1113–1123

Capilla A et al (2012) Planar cell polarity controls directional Notch signaling in the Drosophila leg. Development 139:2584–2593. https://doi.org/10.1242/dev.077446

Casares F, Mann RS (2001) The ground state of the ventral appendage in Drosophila. Science 293:1477–1480. https://doi.org/10.1126/science.1062542

Chu J, Dong PD, Panganiban G (2002) Limb type-specific regulation of bric a brac contributes to morphological diversity. Development 129:695–704

Ciechanska E, Dansereau DA, Svendsen PC, Heslip TR, Brook WJ (2007) dAP-2 and defective proventriculus regulate Serrate and Delta expression in the tarsus of Drosophila melanogaster. Genome 50:693–705. https://doi.org/10.1139/g07-043

Cifuentes FJ, Garcia-Bellido A (1997) Proximo-distal specification in the wing disc of Drosophila by the nubbin gene. Proc Natl Acad Sci U S A 94:11405–11410. https://doi.org/10.1073/pnas.94.21.11405

Citi S, Guerrera D, Spadaro D, Shah J (2014) Epithelial junctions and Rho family GTPases: the zonular signalosome. Small GTPases 5:1–15. https://doi.org/10.4161/21541248.2014.973760

Clark HF et al (1995) Dachsous encodes a member of the cadherin superfamily that controls imaginal disc morphogenesis in Drosophila. Genes Dev 9:1530–1542. https://doi.org/10.1101/gad.9.12.1530

Cohen B, Simcox AA, Cohen SM (1993) Allocation of the thoracic imaginal primordia in the Drosophila embryo. Development 117:597–608

Cordoba S, Estella C (2014) The bHLH-PAS transcription factor dysfusion regulates tarsal joint formation in response to Notch activity during drosophila leg development. PLoS Genet 10:e1004621. https://doi.org/10.1371/journal.pgen.1004621

Cordoba S, Estella C (2018) The transcription factor Dysfusion promotes fold and joint morphogenesis through regulation of Rho1. PLoS Genet 14:e1007584. https://doi.org/10.1371/journal.pgen.1007584

Cordoba S, Requena D, Jory A, Saiz A, Estella C (2016) The evolutionarily conserved transcription factor Sp1 controls appendage growth through Notch signaling. Development 143:3623–3631. https://doi.org/10.1242/dev.138735

Dahmann C, Oates AC, Brand M (2011) Boundary formation and maintenance in tissue development. Nat Rev Genet 12:43–55. https://doi.org/10.1038/nrg2902

Dawes-Hoang RE et al (2005) folded gastrulation, cell shape change and the control of myosin localization. Development 132:4165–4178. https://doi.org/10.1242/dev.01938

de Celis Ibeas JM, Bray SJ (2003) Bowl is required downstream of Notch for elaboration of distal limb patterning. Development 130:5943–5952. https://doi.org/10.1242/dev.00833

de Celis JF, Tyler DM, de Celis J, Bray SJ (1998) Notch signalling mediates segmentation of the Drosophila leg. Development 125:4617–4626

de Madrid BH, Greenberg L, Hatini V (2015) RhoGAP68F controls transport of adhesion proteins in Rab4 endosomes to modulate epithelial morphogenesis of Drosophila leg discs. Dev Biol 399:283–295. https://doi.org/10.1016/j.ydbio.2015.01.004

del Valle Rodriguez A, Didiano D, Desplan C (2012) Power tools for gene expression and clonal analysis in Drosophila. Nature methods 9:47–55. https://doi.org/10.1038/nmeth.1800

Devenport D (2014) The cell biology of planar cell polarity. J Cell Biol 207:171–179. https://doi.org/10.1083/jcb.201408039

Dexter JS (1914) The analysis of a case of continuous variation in Drosophila by a study of its linkage relations. Am Nat 48:712–758. https://doi.org/10.1086/279446

Diaz-Benjumea FJ, Cohen B, Cohen SM (1994) Cell interaction between compartments establishes the proximal-distal axis of Drosophila legs. Nature 372:175–179. https://doi.org/10.1038/372175a0

Duncan DM, Burgess EA, Duncan I (1998) Control of distal antennal identity and tarsal development in Drosophila by spineless-aristapedia, a homolog of the mammalian dioxin receptor. Genes Dev 12:1290–1303

Estella C, Mann RS (2008) Logic of Wg and Dpp induction of distal and medial fates in the Drosophila leg. Development 135:627–636. https://doi.org/10.1242/dev.014670

Estella C, McKay DJ, Mann RS (2008) Molecular integration of wingless, decapentaplegic, and autoregulatory inputs into Distalless during Drosophila leg development. Dev Cell 14:86–96. https://doi.org/10.1016/j.devcel.2007.11.002

Estella C, Voutev R, Mann RS (2012) A dynamic network of morphogens and transcription factors patterns the fly leg. Curr Top Dev Biol 98:173–198. https://doi.org/10.1016/B978-0-12-386499-4.00007-0

Fox DT, Peifer M (2007) Abelson kinase (Abl) and RhoGEF2 regulate actin organization during cell constriction in Drosophila. Development 134:567–578. https://doi.org/10.1242/dev.02748

Fristrom D, Fristrom JW (1993) The metamorphic development of the adult epidermis. In: Bate M, Martinez Arias A (eds) The development of Drosophila melanogaster. Cold Spring Harbor Laboratory Press, Plainview, NY. vol II, pp 843–897

Galindo MI, Bishop SA, Greig S, Couso JP (2002) Leg patterning driven by proximal-distal interactions and EGFR signaling. Science 297:256–259. https://doi.org/10.1126/science.1072311

Galindo MI, Bishop SA, Couso JP (2005) Dynamic EGFR-Ras signalling in Drosophila leg development. Dev Dyn 233:1496–1508. https://doi.org/10.1002/dvdy.20452

Gilmour D, Rembold M, Leptin M (2017) From morphogen to morphogenesis and back. Nature 541:311–320. https://doi.org/10.1038/nature21348

Giorgianni MW, Mann RS (2011) Establishment of medial fates along the proximodistal axis of the Drosophila leg through direct activation of dachshund by Distalless. Dev Cell 20:455–468. https://doi.org/10.1016/j.devcel.2011.03.017

Godt D, Couderc JL, Cramton SE, Laski FA (1993) Pattern formation in the limbs of Drosophila: bric a brac is expressed in both a gradient and a wave-like pattern and is required for specification and proper segmentation of the tarsus. Development 119:799–812

González-Crespo S, Morata G (1996) Genetic evidence for the subdivision of the arthropod limb into coxopodite and telopodite. Development 122:3921–3928

Gonzalez-Crespo S et al (1998) Antagonism between extradenticle function and Hedgehog signalling in the developing limb. Nature 394:196–200. https://doi.org/10.1038/28197

Greenberg L, Hatini V (2009) Essential roles for lines in mediating leg and antennal proximodistal patterning and generating a stable Notch signaling interface at segment borders. Dev Biol 330:93–104. https://doi.org/10.1016/j.ydbio.2009.03.014

Greenberg L, Hatini V (2011) Systematic expression and loss-of-function analysis defines spatially restricted requirements for Drosophila RhoGEFs and RhoGAPs in leg morphogenesis. Mech Dev 128:5–17. https://doi.org/10.1016/j.mod.2010.09.001

Greenwald I (2012) Notch and the awesome power of genetics. Genetics 191:655–669. https://doi.org/10.1534/genetics.112.141812

Grimaldi D, Engel M (2005) Evolution of the insects. Cambridge University Press, New York

Guarner A et al (2014) The zinc finger homeodomain-2 gene of Drosophila controls Notch targets and regulates apoptosis in the tarsal segments. Dev Biol 385:350–365. https://doi.org/10.1016/j.ydbio.2013.10.011

Halder G, Johnson RL (2011) Hippo signaling: growth control and beyond. Development 138:9–22. https://doi.org/10.1242/dev.045500

Hao I, Green RB, Dunaevsky O, Lengyel JA, Rauskolb C (2003) The odd-skipped family of zinc finger genes promotes Drosophila leg segmentation. Dev Biol 263:282–295

Harvey BM, Haltiwanger RS (2018) Regulation of Notch function by O-glycosylation. Adv Exp Med Biol 1066:59–78. https://doi.org/10.1007/978-3-319-89512-3_4

Henrique D, Schweisguth F (2019) Mechanisms of Notch signaling: a simple logic deployed in time and space. Development 146(3):dev172148. https://doi.org/10.1242/dev.172148

Hernandez-Martinez R, Covarrubias L (2011) Interdigital cell death function and regulation: new insights on an old programmed cell death model. Dev Growth Differ 53:245–258. https://doi.org/10.1111/j.1440-169X.2010.01246.x

Homem CC, Peifer M (2008) Diaphanous regulates myosin and adherens junctions to control cell contractility and protrusive behavior during morphogenesis. Development 135:1005–1018. https://doi.org/10.1242/dev.016337

Hori K, Sen A, Artavanis-Tsakonas S (2013) Notch signaling at a glance. J Cell Sci 126:2135–2140. https://doi.org/10.1242/jcs.127308

Jaffe AB, Hall A (2005) Rho GTPases: biochemistry and biology. Annu Rev Cell Dev Biol 21:247–269. https://doi.org/10.1146/annurev.cellbio.21.020604.150721

Johnston LA, Schubiger G (1996) Ectopic expression of wingless in imaginal discs interferes with decapentaplegic expression and alters cell determination. Development 122:3519–3529

Karess RE et al (1991) The regulatory light chain of nonmuscle myosin is encoded by spaghetti-squash, a gene required for cytokinesis in Drosophila. Cell 65:1177–1189

Kerber B, Monge I, Mueller M, Mitchell PJ, Cohen SM (2001) The AP-2 transcription factor is required for joint formation and cell survival in Drosophila leg. Development 128:1231–1238

Kojima T (2017) Developmental mechanism of the tarsus in insect legs. Curr Opin Insect Sci 19:36–42. https://doi.org/10.1016/j.cois.2016.11.002

Kolsch V, Seher T, Fernandez-Ballester GJ, Serrano L, Leptin M (2007) Control of Drosophila gastrulation by apical localization of adherens junctions and RhoGEF2. Science 315:384–386. https://doi.org/10.1126/science.1134833

Kopan R, Ilagan MX (2009) The canonical Notch signaling pathway: unfolding the activation mechanism. Cell 137:216–233. https://doi.org/10.1016/j.cell.2009.03.045

Kozu S et al (2006) Temporal regulation of late expression of Bar homeobox genes during Drosophila leg development by Spineless, a homolog of the mammalian dioxin receptor. Dev Biol 294:497–508. https://doi.org/10.1016/j.ydbio.2006.03.015

Kuhn S, Geyer M (2014) Formins as effector proteins of Rho GTPases. Small GTPases 5:e29513. https://doi.org/10.4161/sgtp.29513

Lai EC (2002) Keeping a good pathway down: transcriptional repression of Notch pathway target genes by CSL proteins. EMBO Rep 3:840–845. https://doi.org/10.1093/embo-reports/kvf170

Lecuit T, Cohen SM (1997) Proximal-distal axis formation in the Drosophila leg. Nature 388:139–145. https://doi.org/10.1038/40563

Leptin M, Grunewald B (1990) Cell shape changes during gastrulation in Drosophila. Development 110:73–84

Liu R, Linardopoulou EV, Osborn GE, Parkhurst SM (2010) Formins in development: orchestrating body plan origami. Biochim Biophys Acta 1803:207–225. https://doi.org/10.1016/j.bbamcr.2008.09.016

Lohmann I, McGinnis N, Bodmer M, McGinnis W (2002) The Drosophila Hox gene deformed sculpts head morphology via direct regulation of the apoptosis activator reaper. Cell 110:457–466

Mack NA, Georgiou M (2014) The interdependence of the Rho GTPases and apicobasal cell polarity. Small GTPases 5:10. https://doi.org/10.4161/21541248.2014.973768

Manjon C, Sanchez-Herrero E, Suzanne M (2007) Sharp boundaries of Dpp signalling trigger local cell death required for Drosophila leg morphogenesis. Nat Cell Biol 9:57–63. https://doi.org/10.1038/ncb1518

Mao Y et al (2006) Dachs: an unconventional myosin that functions downstream of Fat to regulate growth, affinity and gene expression in Drosophila. Development 133:2539–2551. https://doi.org/10.1242/dev.02427

Marston DJ et al (2016) MRCK-1 drives apical constriction in C. elegans by linking developmental patterning to force generation. Curr Biol 26:2079–2089. https://doi.org/10.1016/j.cub.2016.06.010

Martin AC, Goldstein B (2014) Apical constriction: themes and variations on a cellular mechanism driving morphogenesis. Development 141:1987–1998. https://doi.org/10.1242/dev.102228

Martin AC, Kaschube M, Wieschaus EF (2009) Pulsed contractions of an actin-myosin network drive apical constriction. Nature 457:495–499. https://doi.org/10.1038/nature07522

Martin AC, Gelbart M, Fernandez-Gonzalez R, Kaschube M, Wieschaus EF (2010) Integration of contractile forces during tissue invagination. J Cell Biol 188:735–749. https://doi.org/10.1083/jcb.200910099

Mason FM, Tworoger M, Martin AC (2013) Apical domain polarization localizes actin-myosin activity to drive ratchet-like apical constriction. Nat Cell Biol 15:926–936. https://doi.org/10.1038/ncb2796

Mason FM, Xie S, Vasquez CG, Tworoger M, Martin AC (2016) RhoA GTPase inhibition organizes contraction during epithelial morphogenesis. J Cell Biol 214:603–617. https://doi.org/10.1083/jcb.201603077

McKay DJ, Estella C, Mann RS (2009) The origins of the Drosophila leg revealed by the cis-regulatory architecture of the Distalless gene. Development 136:61–71. https://doi.org/10.1242/dev.029975

Mirth C, Akam M (2002) Joint development in the Drosophila leg: cell movements and cell populations. Dev Biol 246:391–406. https://doi.org/10.1006/dbio.2002.0593

Monge I et al (2001) Drosophila transcription factor AP-2 in proboscis, leg and brain central complex development. Development 128:1239–1252

Monier B, Suzanne M (2015) The morphogenetic role of apoptosis. Curr Top Dev Biol 114:335–362. https://doi.org/10.1016/bs.ctdb.2015.07.027

Monier B et al (2015) Apico-basal forces exerted by apoptotic cells drive epithelium folding. Nature 518:245–248. https://doi.org/10.1038/nature14152

Morgan TH (1917) The theory of the gene. Am Nat 51:513–544. https://doi.org/10.1086/279629

Morimura S, Maves L, Chen Y, Hoffmann FM (1996) decapentaplegic overexpression affects Drosophila wing and leg imaginal disc development and wingless expression. Dev Biol 177:136–151. https://doi.org/10.1006/dbio.1996.0151

Mulinari S, Barmchi MP, Hacker U (2008) DRhoGEF2 and diaphanous regulate contractile force during segmental groove morphogenesis in the Drosophila embryo. Mol Biol Cell 19:1883–1892. https://doi.org/10.1091/mbc.E07-12-1230

Munoz-Descalzo S et al (2010) Wingless modulates the ligand independent traffic of Notch through Dishevelled. Fly (Austin) 4:182–193. https://doi.org/10.4161/fly.4.3.11998

Natori K, Tajiri R, Furukawa S, Kojima T (2012) Progressive tarsal patterning in the Drosophila by temporally dynamic regulation of transcription factor genes. Dev Biol 361:450–462. https://doi.org/10.1016/j.ydbio.2011.10.031

Neisch AL, Speck O, Stronach B, Fehon RG (2010) Rho1 regulates apoptosis via activation of the JNK signaling pathway at the plasma membrane. J Cell Biol 189:311–323. https://doi.org/10.1083/jcb.200912010

Neubueser D, Hipfner DR (2010) Overlapping roles of Drosophila Drak and Rok kinases in epithelial tissue morphogenesis. Mol Biol Cell 21:2869–2879. https://doi.org/10.1091/mbc.E10-04-0328

Newcomb S et al (2018) cis-regulatory architecture of a short-range EGFR organizing center in the Drosophila melanogaster leg. PLoS Genet 14:e1007568. https://doi.org/10.1371/journal.pgen.1007568

Ng M, Diaz-Benjumea FJ, Cohen SM (1995) Nubbin encodes a POU-domain protein required for proximal-distal patterning in the Drosophila wing. Development 121:589–599

Nusslein-Volhard C, Wieschaus E (1980) Mutations affecting segment number and polarity in Drosophila. Nature 287:795–801

Okajima T, Irvine KD (2002) Regulation of notch signaling by O-linked fucose. Cell 111:893–904. https://doi.org/10.1016/S0092-8674(02)01114-5

Panin VM, Papayannopoulos V, Wilson R, Irvine KD (1997) Fringe modulates Notch-ligand interactions. Nature 387:908–912. https://doi.org/10.1038/43191

Pastor-Pareja JC, Grawe F, Martin-Blanco E, Garcia-Bellido A (2004) Invasive cell behavior during Drosophila imaginal disc eversion is mediated by the JNK signaling cascade. Dev Cell 7:387–399. https://doi.org/10.1016/j.devcel.2004.07.022

Penton AL, Leonard LD, Spinner NB (2012) Notch signaling in human development and disease. Semin Cell Dev Biol 23:450–457. https://doi.org/10.1016/j.semcdb.2012.01.010

Pueyo JI, Couso JP (2008) The 11-aminoacid long Tarsal-less peptides trigger a cell signal in Drosophila leg development. Dev Biol 324:192–201. https://doi.org/10.1016/j.ydbio.2008.08.025

Pueyo JI, Couso JP (2011) Tarsal-less peptides control Notch signalling through the Shavenbaby transcription factor. Dev Biol 355:183–193. https://doi.org/10.1016/j.ydbio.2011.03.033

Pueyo JI, Galindo MI, Bishop SA, Couso JP (2000) Proximal-distal leg development in Drosophila requires the apterous gene and the Lim1 homologue dlim1. Development 127:5391–5402

Rauskolb C (2001) The establishment of segmentation in the Drosophila leg. Development 128:4511–4521

Rauskolb C, Irvine KD (1999) Notch-mediated segmentation and growth control of the Drosophila leg. Dev Biol 210:339–350. https://doi.org/10.1006/dbio.1999.9273

Rauskolb C, Correia T, Irvine KD (1999) Fringe-dependent separation of dorsal and ventral cells in the Drosophila wing. Nature 401:476–480. https://doi.org/10.1038/46786

Reeves GT, Stathopoulos A (2009) Graded dorsal and differential gene regulation in the Drosophila embryo. Cold Spring Harb Perspect Biol 1:a000836. https://doi.org/10.1101/cshperspect.a000836

Riento K, Ridley AJ (2003) Rocks: multifunctional kinases in cell behaviour. Nat Rev Mol Cell Biol 4:446–456. https://doi.org/10.1038/nrm1128

Robertson F, Pinal N, Fichelson P, Pichaud F (2012) Atonal and EGFR signalling orchestrate rok- and Drak-dependent adherens junction remodelling during ommatidia morphogenesis. Development 139:3432–3441. https://doi.org/10.1242/dev.080762

Roh-Johnson M et al (2012) Triggering a cell shape change by exploiting preexisting actomyosin contractions. Science 335:1232–1235. https://doi.org/10.1126/science.1217869

Rosenblatt J, Raff MC, Cramer LP (2001) An epithelial cell destined for apoptosis signals its neighbors to extrude it by an actin- and myosin-dependent mechanism. Curr Biol 11:1847–1857

Ruiz-Losada M, Blom-Dahl D, Cordoba S, Estella C (2018) Specification and patterning of drosophila appendages. J Dev Biol 6(3):17. https://doi.org/10.3390/jdb6030017

Sawyer JM et al (2010) Apical constriction: a cell shape change that can drive morphogenesis. Dev Biol 341:5–19. https://doi.org/10.1016/j.ydbio.2009.09.009

Shirai T, Yorimitsu T, Kiritooshi N, Matsuzaki F, Nakagoshi H (2007) Notch signaling relieves the joint-suppressive activity of Defective proventriculus in the Drosophila leg. Dev Biol 312:147–156. https://doi.org/10.1016/j.ydbio.2007.09.003

Shubin N, Tabin C, Carroll S (1997) Fossils, genes and the evolution of animal limbs. Nature 388:639–648. https://doi.org/10.1038/41710

Simoes S et al (2006) Compartmentalisation of Rho regulators directs cell invagination during tissue morphogenesis. Development 133:4257–4267. https://doi.org/10.1242/dev.02588

Slattum G, McGee KM, Rosenblatt J (2009) P115 RhoGEF and microtubules decide the direction apoptotic cells extrude from an epithelium. J Cell Biol 186:693–702. https://doi.org/10.1083/jcb.200903079

Snodgrass RE (1935) Principles of insect morphology. McGraw Hill, New York, 667 pp

Soler C, Daczewska M, Da Ponte JP, Dastugue B, Jagla K (2004) Coordinated development of muscles and tendons of the Drosophila leg. Development 131:6041–6051. https://doi.org/10.1242/dev.01527

Struhl G, Basler K (1993) Organizing activity of wingless protein in Drosophila. Cell 72:527–540

Suzanne M (2016) Molecular and cellular mechanisms involved in leg joint morphogenesis. Semin Cell Dev Biol 55:131–138. https://doi.org/10.1016/j.semcdb.2016.01.032

Suzanne M, Steller H (2013) Shaping organisms with apoptosis. Cell Death Differ 20:669–675. https://doi.org/10.1038/cdd.2013.11

Suzanne M et al (2010) Coupling of apoptosis and L/R patterning controls stepwise organ looping. Curr Biol 20:1773–1778. https://doi.org/10.1016/j.cub.2010.08.056

Svendsen PC, Ryu JR, Brook WJ (2015) The expression of the T-box selector gene midline in the leg imaginal disc is controlled by both transcriptional regulation and cell lineage. Biol Open 4:1707–1714. https://doi.org/10.1242/bio.013565

Tajiri R, Misaki K, Yonemura S, Hayashi S (2010) Dynamic shape changes of ECM-producing cells drive morphogenesis of ball-and-socket joints in the fly leg. Development 137:2055–2063. https://doi.org/10.1242/dev.047175

Tajiri R, Misaki K, Yonemura S, Hayashi S (2011) Joint morphology in the insect leg: evolutionary history inferred from Notch loss-of-function phenotypes in Drosophila. Development 138:4621–4626. https://doi.org/10.1242/dev.067330

Tan JL, Ravid S, Spudich JA (1992) Control of nonmuscle myosins by phosphorylation. Annu Rev Biochem 61:721–759. https://doi.org/10.1146/annurev.bi.61.070192.003445

Theisen H, Haerry TE, O'Connor MB, Marsh JL (1996) Developmental territories created by mutual antagonism between Wingless and Decapentaplegic. Development 122:3939–3948

Vasquez CG, Martin AC (2016) Force transmission in epithelial tissues. Dev Dyn 245:361–371. https://doi.org/10.1002/dvdy.24384

Vidal M, Larson DE, Cagan RL (2006) Csk-deficient boundary cells are eliminated from normal Drosophila epithelia by exclusion, migration, and apoptosis. Dev Cell 10:33–44. https://doi.org/10.1016/j.devcel.2005.11.007

Villano JL, Katz FN (1995) 4-jointed is required for intermediate growth in the proximal-distal Axis in Drosophila. Development 121:2767–2777

von Kalm L, Fristrom D, Fristrom J (1995) The making of a fly leg: a model for epithelial morphogenesis. BioEssays 17:693–702. https://doi.org/10.1002/bies.950170806

Weng M, Wieschaus E (2016) Myosin-dependent remodeling of adherens junctions protects junctions from Snail-dependent disassembly. J Cell Biol 212:219–229. https://doi.org/10.1083/jcb.201508056

Wilder EL, Perrimon N (1995) Dual functions of wingless in the Drosophila leg imaginal disc. Development 121:477–488

Winter CG et al (2001) Drosophila Rho-associated kinase (Drok) links Frizzled-mediated planar cell polarity signaling to the actin cytoskeleton. Cell 105:81–91

Xu N, Keung B, Myat MM (2008) Rho GTPase controls invagination and cohesive migration of the Drosophila salivary gland through Crumbs and Rho-kinase. Dev Biol 321:88–100. https://doi.org/10.1016/j.ydbio.2008.06.007

Yamaguchi Y et al (2011) Live imaging of apoptosis in a novel transgenic mouse highlights its role in neural tube closure. J Cell Biol 195:1047–1060. https://doi.org/10.1083/jcb.201104057

Zegers MM, Friedl P (2014) Rho GTPases in collective cell migration. Small GTPases 5:e28997. https://doi.org/10.4161/sgtp.28997

Chapter 8
Notch Signalling: The Multitask Manager of Inner Ear Development and Regeneration

Nicolas Daudet and Magdalena Żak

Abstract Notch signalling is a major regulator of cell fate decisions and tissue patterning in metazoans. It is best known for its role in lateral inhibition, whereby Notch mediates competitive interactions between cells to limit adoption of a given developmental fate. However, it can also function by lateral induction, a cooperative mode of action that was originally described during the patterning of the *Drosophila* wing disc and creates boundaries or domains of cells of the same character. In this chapter, we introduce these two signalling modes and explain how they contribute to distinct aspects of the development and regeneration of the vertebrate inner ear, the organ responsible for the perception of sound and head movements. We discuss some of the factors that could influence the context-specific outcomes of Notch signalling in the inner ear and the ongoing efforts to target this pathway for the treatment of hearing loss and vestibular dysfunction.

Keywords Notch signalling · Lateral inhibition · Lateral induction · Proneural genes · Cell fate decisions · Inner ear · Cochlea · Organ of Corti · Development · Prosensory specification · Hair cell · Deafness · Hair cell regeneration

Introduction

A great diversity of fate decisions and cellular processes are regulated by Notch signalling, due to context-dependent differences in its transcriptional targets and the multitude of factors that influence the spatial pattern and dynamics of Notch activity. The development of the inner ear provides a great illustration of this versatility. Notch controls several key cell determination and patterning events during the differentiation of the neurosensory cells of the inner ear, through different ligands and modes of signalling. It is critical for the formation of the mechanosensory 'hair'

N. Daudet (✉) · M. Żak
University College London, The Ear Institute, London, UK
e-mail: n.daudet@ucl.ac.uk

J. Reichrath, S. Reichrath (eds.), *Notch Signaling in Embryology and Cancer*, Advances in Experimental Medicine and Biology 1218,
https://doi.org/10.1007/978-3-030-34436-8_8

cells (HCs) that populate the sensory organs of the inner ear and are essential for hearing and our perception of balance. This has prompted considerable interest in the potential manipulation of Notch activity to stimulate HC regeneration in the damaged ear, a topic discussed at the end of this review.

The Notch Signalling Pathway

We provide below a very brief overview of the mechanisms of Notch signalling, lateral inhibition and lateral induction. For additional molecular and biochemical details, or non-canonical modes of action of Notch, we refer the reader to other reviews (Bray 2006; Henrique and Schweisguth 2019; Yamamoto et al. 2010).

Basic Components of the Canonical Notch Pathway Notch receptors (Fig. 8.1) are transmembrane proteins, composed of an extracellular domain with multiple EGF-like repeats and a Notch intracellular domain (NICD). The Notch receptors are activated by transmembrane ligands belonging to the Delta/Serrate/Lag-2 (DSL) family. Binding of the DSL ligands to the Notch extracellular domain, however, is not sufficient for receptor activation. Their internalization in the 'signal-sending' cell, which is regulated by the E3-ubiquitin ligases of the Mindbomb and Neuralized families, is required to expose two proteolytic sites near the Notch transmembrane domain. These are then cleaved by two metalloproteases [protease A-disintegrin

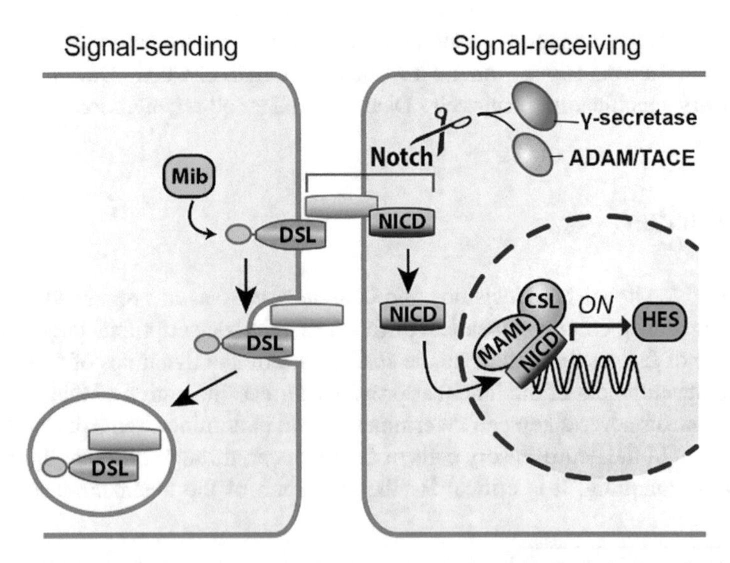

Fig. 8.1 Notch signalling. The endocytosis of the DSL ligand, which depends on Mib/Neur activity, triggers the proteolytic cleavage of the Notch receptor in the signal-receiving cell. The Notch extracellular domain (in green) is internalized with the DSL ligand, whilst the NICD translocates to the nucleus and activates the expression of specific transcriptional target, such as those of the Hes family

and metalloprotease-10 (ADAM10) and tumour necrosis factor alpha converting enzyme (TACE)] and the γ-secretase enzyme. The NICD is then released in the 'signal-receiving' cell, whilst the Notch extracellular domain is endocytosed with the ligand in the signal-sending cell. Following its cleavage, NICD is translocated to the nucleus where it forms a transcriptional complex with CSL (CBF1/SuH/Lag-1) and Mastermind-like (MAML) to activate the expression of Notch target genes. The CSL protein can also function as a transcriptional repressor in the absence of NICD, keeping some of the Notch targets silenced. The classic direct targets and effectors of Notch belong to the *hairy and enhancer of split (Hes)* and *Hes-related* gene family, which encode basic helix-loop-helix (bHLH) transcriptional repressors. They antagonize the expression and activity of other genes, in particular the proneural bHLH factors.

Two Contrasting Modes of Signalling: Lateral Inhibition and Lateral Induction Importantly, Notch activity can feedback positively or negatively on the expression of the DSL ligands. This produces two distinct modes of signalling, lateral inhibition and lateral induction, which have different outcomes in terms of cell differentiation and patterning for interacting cells.

	Lateral inhibition	Lateral induction
Starting conditions	Equivalence group	Context-dependent
Target of notch activity	Proneural factors	Context-dependent
Regulation of DSL expression by notch activity	Negative	Positive
Cellular outcome	Alternate fates	Same fate

In lateral inhibition (Fig. 8.2a), interacting cells compete for the adoption of the primary fate, defined by the expression of a proneural bHLH factor that (i) promotes DSL ligand expression and (ii) is repressed by Notch activity. Starting from a condition where all cells are (in theory) equal in their developmental potential, random variations in the expression of the proneural factor lead some cells to elevate their DSL levels – these become better signal-sending cells. The signal-receiving cells, on the other hand, reduce their levels of proneural factor and DSL expression; they are consequently diverted from the primary fate. The outcome of lateral inhibition is a diversification of cell fates, creating a branching point within a cell lineage, or a salt-and-pepper mosaic of two cell types within an epithelium. Its failure results in the overproduction of cells adopting the primary fate. Lateral inhibition is a very common and evolutionary conserved mechanism regulating, for example, cell diversification in the sensory organs of Drosophila, neurogenesis, the formation of secretory cells in the gut or the production of HCs in the inner ear.

In lateral induction (Fig. 8.2b), interacting cells cooperate to maintain Notch activity and adopt the same fate. Lateral induction is not as common as lateral inhibition, but it has been well studied in the *Drosophila* wing disc (see later), and it regulates vascular smooth muscle differentiation (Manderfield et al. 2012), neural crest induction (Cornell and Eisen 2005), lens fibre differentiation (Saravanamuthu

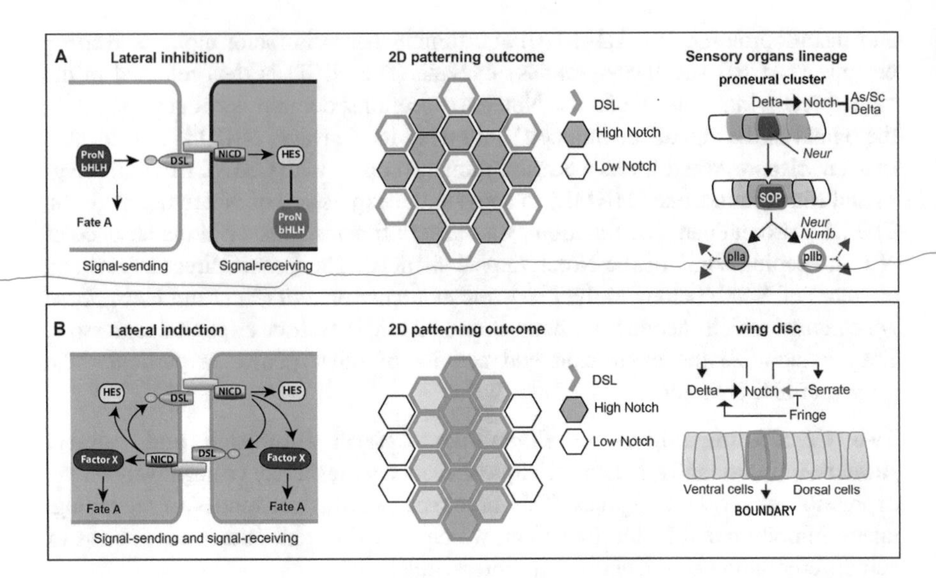

Fig. 8.2 Two contrasting modes of Notch signalling. (**a**) In lateral inhibition, a proneural bHLH factor induces DSL expression, and its expression is repressed by Notch in the signal-receiving cell. This results in a diversification of cell fates or the formation of a salt-and-pepper cellular mosaic. In the proneural clusters giving rise to the adult mechanosensory bristles of Drosophila, the proneural bHLH factor Achaete-Scute (As/Sc) is restricted by lateral inhibition to the sensory organ precursor cell (SOP). The SOP gains a strong advantage due to the expression of Neuralized (Neur). Lateral inhibition also operates in the progeny of the SOP and is biased by the differential distribution of Neur and Numb in one of the daughter cells after each round of cell division. (**b**) In lateral induction, interacting cells are at the same time signal-receiving and signal-sending and adopt the same character. In the Drosophila wing disc, Delta and Serrate are regulated by lateral induction; Fringe, which is expressed in the dorsal compartment, enables strong Notch activation at the dorsoventral boundary by reducing Serrate1/Notch and increasing Delta1/Notch signalling. In all diagrams, the arrows do not necessarily imply direct regulation or interaction

et al. 2009), interactions between the epidermis and the dermis (Ambler and Watt 2010) and prosensory specification in the inner ear.

Cells with an Edge: Factors Improving Lateral Inhibition A critical parameter for efficient lateral inhibition is the strength of the intercellular feedback loop repressing the DSL ligand in the signal-receiving cells. However, signal-sending cells can use several tricks to improve their chances of delivering a 'loud and clear' message to their neighbours. For example, in Drosophila proneural clusters (Fig. 8.2a), Neuralized is restricted to the nascent sensory organ precursor (SOP), which enables these to deliver efficient lateral inhibition to neighbouring cells (Yamamoto et al. 2010). In the following rounds of division, lateral inhibition is also biased by the asymmetric inheritance of Neuralized and the endocytotic protein Numb, which reduces cell-surface levels of Notch receptors in one of the daughter cells (Couturier et al. 2013). The DSL ligands can also bind to Notch receptors in 'cis' (within the signal-sending cell) to prevent their activation (del Álamo et al. 2011). Finally, the expression of dominant-negative HLH (*emc* in *Drosophila*, *inhibitor of differentiation* or *Id* genes in vertebrates) can restrict the activity of the proneural bHLH genes to specific cells, which gain a strong competitive advantage in the race for adoption of the primary fate (Troost et al. 2015).

Hence, more than the expression levels of a DSL ligand, Notch receptor or proneural gene, it is the *activity levels* of these components that determine, in a context-dependent manner, the outcomes of lateral inhibition.

Cells at the Edge: Lateral Induction and the Making of Tissue Boundaries The best-characterized example of lateral induction occurs during the formation of the dorsoventral lineage boundary of the *Drosophila* wing (imaginal) disc (Fig. 8.2b), which in the adult gives rise to the peripheral wing margin. The DV boundary acts as a cellular fence, which prevents cells belonging to the dorsal 'compartment' of the wing disc from mixing with those of the ventral compartment during tissue growth (reviewed in Dahmann et al. 2011). Notch signalling is necessary for the formation of the boundary. Each compartment expresses uniformly a DSL ligand (Serrate in the dorsal compartment, Delta in the ventral one), and their expression is stimulated by Notch activity (de Celis and Bray 1997). But although the Notch receptor is present throughout the wing disc, Notch activity is strongly elicited along the DV boundary only, in a one- to three-cell-wide domain where expression of the ligands also becomes elevated (de Celis et al. 1996). This restricted activation is due to the action of Fringe, a glycosyltransferase present in the dorsal compartment that modifies the extracellular domain of the Notch receptor to make it less sensitive to Serrate, but promotes Delta/Notch interactions in both cis- and trans-signalling (LeBon et al. 2014; Rauskolb et al. 1999). Consequently, Delta (in ventral cells) can strongly activate Notch in the (dorsal) Fringe-expressing cells – creating a longitudinal band of Notch activity between the dorsal and ventral compartments.

Hence, the differing modes of regulation of DSL ligands and the modification of Notch receptors can radically transform the tissue-level outcomes of Notch signalling. This versatility is also manifest in the inner ear, where Notch signalling is truly multi-tasking and managing through different ligands and modes of action, some of the key aspects of its development.

Introduction to the Inner Ear

Aptly named by the early anatomists the 'labyrinth', the membranous part of the inner ear is composed of a complex 3D network of fluid-filled canals and chambers, lined up by specialized epithelial cells. The mammalian inner ear can be subdivided into a dorsal part containing the vestibular system and a ventral part, the cochlea (Fig. 8.3). The vestibular system has three semi-circular canals oriented along orthogonal axes and connected to three sensory organs, called cristae, which are sensitive to the angular rotation of the head. In addition, it contains the maculae of the utricle and the saccule, which act as gravity and acceleration sensors. In mammals, the cochlea forms a coiled structure resembling a snail shell and hosts a sensory

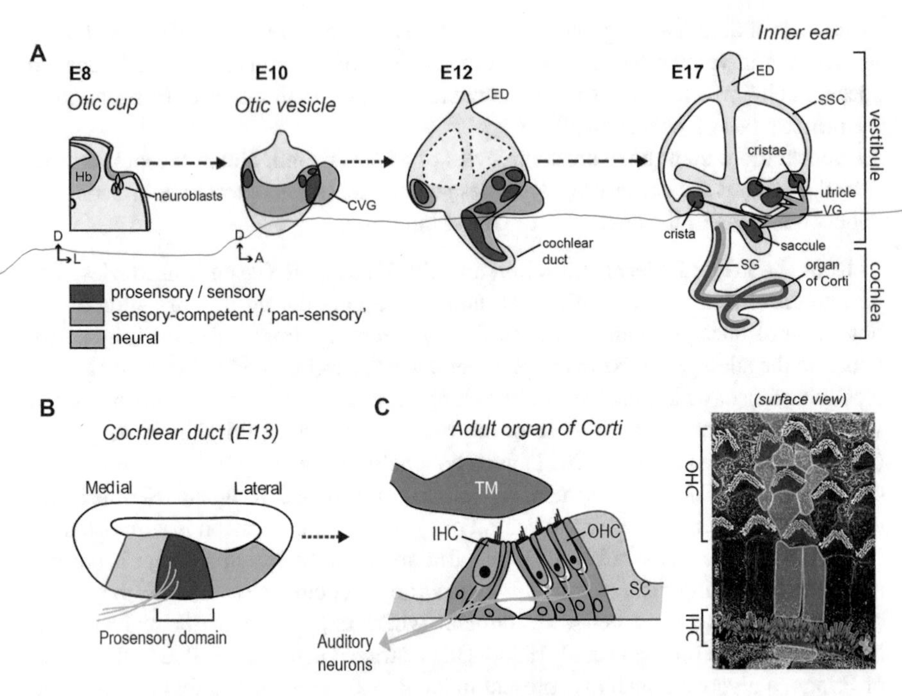

Fig. 8.3 Development of the mouse inner ear. (**a**) The otic cup invaginates in the underlying mesenchyme and closes into a vesicle, which then gives rise to the different structures of the inner ear. The neuroblasts delaminate from the anterior part of the otic placode/cup to form the CVG ganglion, from which the vestibular and auditory neurons innervating the HCs derive. The specification of the sensory organs is coupled to their progressive segregation from a large sensory-competent domain. By E17, the inner ear has reached an adult-like morphology. (**b**, **c**) Transverse representation of the embryonic cochlear duct, with its prosensory domain, from which the adult SCs and HCs of the organ of Corti derive. The IHCs are the main sensory transducers and connected by the majority of nerve afferents, whilst the OHCs have electromotile properties essential for the sensitivity and frequency selectivity of the cochlea. Both types of HCs are interspaced by the cell bodies and apical surfaces of different types of specialized SCs. (**d**) Scanning electron microscopy (courtesy of Andy Forge) view of the surface of the organ of Corti and its regular mosaic of HCs (purple) and SCs (green). Abbreviations: E embryonic day, Hb hindbrain, CVG cochleo-vestibular ganglion, ED endolymphatic duct, SSC semi-circular canal, VG vestibular ganglion, SG spiral ganglion, SC supporting cells, IHC inner hair cells, OHC outer hair cells, TM tectorial membrane

epithelium activated by sound, called the organ of Corti (Fig. 8.3b, c). The semicircular canal system is highly conserved, but there are variations in the number and morphology of the other sensory organs across vertebrates. However, one universal feature is the 'salt-and-pepper' mosaic of mechanosensory HCs, interspaced from one another by supporting cells (SCs). The HCs are topped by an array of modified microvilli, or stereocilia, arranged in neat rows forming a staircase-like pattern. The stereociliary bundles are in contact with specialized extracellular gels or membranes and are bathed in a potassium-rich fluid called the endolymph. When inner ear fluids are displaced in response to changes in head position or to the vibrations of the middle ear ossicles, mechanotransduction channels located at the top of the stereocilia open, leading to an influx of potassium ions and cell depolarisation. This causes neurotransmitter release at the synaptic pole of HCs and the stimulation of the afferent neurites of the auditory and vestibular neurons, which relay this information to the central nervous system.

Inner Ear Development in a Nutshell We provide here a very brief summary of inner ear development in the mouse (Fig. 8.3) and refer to other reviews for further details (Alsina and Whitfield 2017; Basch et al. 2016a; Fritzsch and Beisel 2001).

The majority of the cells that compose the inner ear (including the audio-vestibular neurons) derive from the otic placode, an epithelial thickening of the head ectoderm located on both sides of the embryonic hindbrain. The placode invaginates and then pinches off the surface ectoderm to form the otic vesicle or otocyst. As it grows, this simple sphere undergoes a drastic remodelling to give rise to various vestibular structures dorsally (endolymphatic duct, semi-circular canals) and the cochlear duct ventrally. This is accompanied by dynamic changes in the expression of molecular factors regulating the specification of otic progenitors. The precursors for the sensory organs are located within *prosensory* domains, which are produced sequentially by segregation from a large sensory-competent domain that extends along the ventromedial wall of the otic vesicle. The prosensory cells then gradually exit the cell cycle and differentiate into HCs and SCs. The first HCs are formed in the vestibular patches. In the embryonic organ of Corti, the terminal mitosis of the prosensory cells proceeds from the distal end to the base of the cochlear duct (around E13–E15 in the mouse), but HC differentiation and other aspects of the maturation of the epithelium follow the opposite direction. The onset of function of the mouse cochlea occurs at approximately 2 weeks of age.

Roles of Notch During Inner Ear Development

In 1991, as the roles of Notch and lateral inhibition were uncovered in the nervous system of *Drosophila* and vertebrates, Julian Lewis proposed that the same mechanism could control the production of the neurosensory cells of the inner ear (Lewis 1991). Since then, the experimental evidence supporting this idea has accumulated, and new roles for Notch have been uncovered in prosensory specification and otic induction.

Lateral Inhibition and the Regulation of Otic Neurogenesis

Starting from approximately E8 and until E12 in the mouse, the precursor cells for the neurons of the cochleo-vestibular ganglion, or neuroblasts, delaminate from the antero-medial domain of the otic placode (then vesicle, see Fig. 8.3). The neurogenic domain is included within a larger neurosensory competent domain,

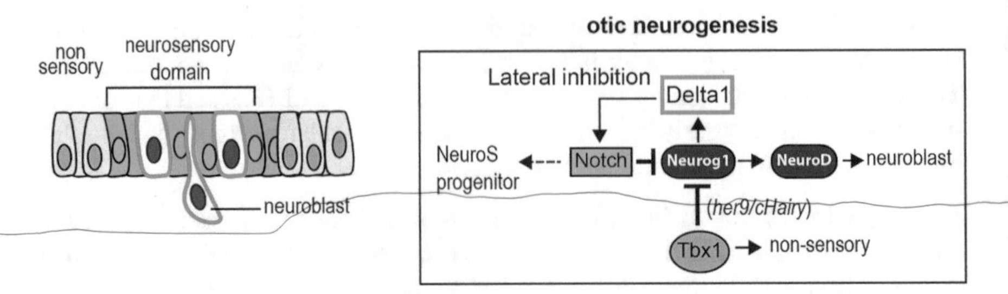

Fig. 8.4 Lateral inhibition during otic neurogenesis. The neuroblasts (cells in white on the left panel) delaminate from the neurosensory domain of the otic placode/vesicle and express Neurog1 and NeuroD. The signal-receiving cells remain as neurosensory competent progenitors. Tbx1 and some effectors of the Notch pathway antagonize Neurog1 outside of the neurosensory domain. The color codes used on the simplified regulatory network (on the right) match those of the drawing (on the left)

which gives rise to several of the sensory organs. After their delamination, neuroblasts proliferate and differentiate into two populations of neurons: the vestibular neurons, which are born first, and the auditory neurons (Koundakjian et al. 2007). The selection of neuroblasts is regulated by lateral inhibition (Fig. 8.4). *Notch1* is present throughout the otic epithelium, but Dll1 is restricted to the neurogenic patch, where it is expressed in a scattered manner (Alsina et al. 2004; Daudet et al. 2007). The neuroblasts express two proneural bHLH factors, Neurogenin1 (Neurog1) and NeuroD, which are, respectively, required for their initial specification and their delamination and survival (Kim et al. 2001; Liu et al. 2000; Ma et al. 2000; Matei et al. 2005). Notch effectors and the modulator *Lunatic Fringe* (*Lfng*) are also expressed there (Adam et al. 1998; Cole et al. 2000). The evidence that lateral inhibition controls neuroblast formation came first from the *mindbomb* zebrafish mutant, which shows excessive neuronal production throughout its nervous system and in the inner ear (Haddon et al. 1998). In the chick otocyst, the pharmacological inhibition of Notch activity with a gamma-secretase inhibitor (GSI) or through overexpression of a dominant-negative form of Mastermind also leads to excess neuronal differentiation and an increase in *Dl1* expression (Abelló et al. 2007; Daudet et al. 2007) as predicted by the standard model of lateral inhibition with feedback.

The factors that establish otic neural competence are still unclear but involve a complex interplay of diffusible signals emanating from the surrounding tissues and transcription factors (reviewed in Gálvez et al. 2017; Raft and Groves 2014). Some effectors of the Notch pathway could also play a part in this process. In fact, *cHairy1* (an orthologue of *Hes1*) in the chick and *her9* in the zebrafish (Radosevic et al. 2011) are expressed outside of the neurogenic domain. However, neither activation of canonical Notch nor a particular ligand has been firmly associated with their regulation. Instead, the transcription factor Tbx1 and retinoic acid, which promotes posterior identity in the otocyst, act upstream of *her9* in zebrafish. The inactivation of *her9* by morpholinos, similar to the absence of retinoic acid or *Tbx1* in the mouse (Raft 2004), leads to ectopic induction of neurogenesis in posterior regions of the otocyst (Radosevic et al. 2011).

Lateral Inhibition and Hair Cell Fate Decisions

Once prosensory cells exit the cell cycle, they differentiate into HCs or SCs, and this decision is controlled by lateral inhibition (Fig. 8.5). Lateral inhibition has been most studied in the organ of Corti, partly because defects in the number and organization of inner HCs and outer HCs, organized, respectively, in one and three parallel rows, are very easy to spot. The basic rules of lateral inhibition appear nevertheless conserved across all inner ear sensory epithelia. The nascent HCs deliver lateral inhibition by expressing, in a transient manner, multiple DSL ligands: Dll1, Jag2 and Dll3 in the mouse (Hartman et al. 2007; Lanford et al. 1999; Morrison et al. 1999), DeltaA and DeltaB in the fish (Haddon et al. 1998; Riley et al. 1999) and at least Dll1 in the chick (Adam et al. 1998). In the signal-receiving cells, Notch activity represses the expression of *Atonal-homologue 1* (*Atoh1* in mammals, *cath1* in the chick, *atoh1a/b* in the fish), a bHLH proneural gene required for the formation of chordotonal organs in the fly, and HCs as well as other cell types (granule cells in the cerebellum, secretory cells of the gut lining, etc.) in vertebrates (Bermingham et al. 1999; reviewed in Jarman and Groves 2013). In zebrafish, *atoh1a* is expressed in a large territory before becoming restricted to the first HCs, suggesting that it defines a genuine equivalence group (Millimaki et al. 2007). In the mouse cochlea, however, *Atoh1* is highly expressed in nascent HCs but much harder to detect in the prosensory cells (Bermingham et al. 1999; Cai et al. 2013; Chen et al. 2002; Driver et al. 2013; Lanford et al. 2000; Woods et al. 2004; Yang et al. 2010), hinting at a different mode of regulation.

Redundancies in the Lateral Inhibition Machinery Ensure Robust Cell Fate Decisions In the *mindbomb* zebrafish, HCs are produced early and in excess at the expense of the SCs (Haddon et al. 1998). In the absence of SCs, these HCs do not survive long and are rapidly eliminated from the epithelium. This remains to date the most dramatic phenotype observed in any Notch mutant, presumably because

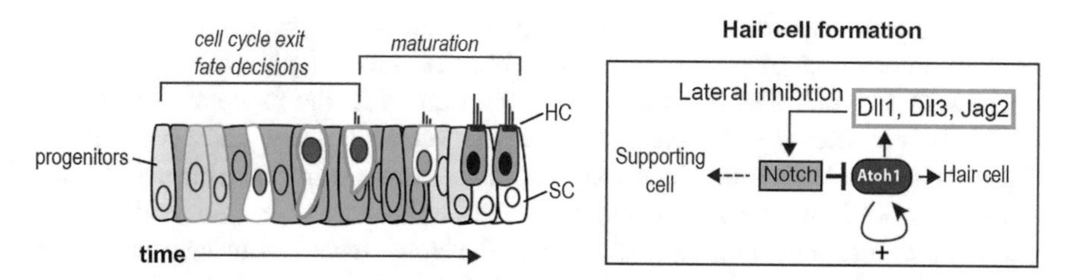

Fig. 8.5 Lateral inhibition during hair cell formation. Left panel: schematic representation of the changes in the expression of Atoh1 (in green), DSL ligands (orange) and levels of Notch activity (blue) in the course of HC formation are represented. After cell cycle exit, nascent HCs upregulate Atoh1 and several DSL, driving high Notch activity in prosensory domains. Following commitment to the HC fate, cells down-regulate Atoh1 and DSL expression; Notch activity decreases in the mature sensory epithelium. The right panel represents the basic regulatory circuit during the lateral inhibition of HC formation. The autoregulatory feedback loop controlling Atoh1 could lead to a rapid elevation of DSL expression levels, ensuring robust lateral inhibition

the E3-ubiquitin ligase Mib is required for the activities of several DSL ligands (Itoh et al. 2003). An overproduction of HCs can also be elicited in organotypic cultures of embryonic organ of Corti by GSI treatment (Tang et al. 2006; Yamamoto et al. 2006), although high doses are required to achieve the strongest phenotype (Doetzlhofer et al. 2009). Whilst this confirms that Notch signalling is a key regulator of HC formation, it also indicates that multiple ligands or receptors must mediate lateral inhibition. There is in fact good evidence that the DSL ligands of HCs act in a cooperative and partly redundant manner: in the mouse organ of Corti of the Jag2 mutant, an additional row of IHCs is present, but the OHCs are unaffected (Lanford et al. 1999). In the *Dll1* conditional Knock-Out (cKO) (Brooker 2006) or hypomorph mutant (Kiernan 2005), both IHCs and OHCs are produced early and in excess, but this phenotype becomes much more dramatic in a compound *Dll1/Jag2* mutant, suggesting synergistic effects (Kiernan 2005). The *Dll3* mutant does not exhibit any defect in HC numbers (Hartman et al. 2007), which suggests either that this ligand is not contributing to lateral inhibition or that its absence is entirely compensated by Dll1 and Jag2. The *Notch1* cKO mouse has a phenotype that is as severe as the combined loss of Jag2 and Dll1 (Kiernan 2005), suggesting that it is the main mediator of lateral inhibition. Its paralogues *Notch2* and *Notch3* are also expressed in the developing inner ear (Basch et al. 2011; Hao et al. 2012; Lindsell et al. 1996; Maass et al. 2015; Yamamoto et al. 2006), but their functions have not been tested. On the other hand, multiple Notch effectors of the *Hes/Hey* family are present in prosensory cells and SCs of the organ of Corti, and these interact genetically: compound mutants for *Hes1*, *Hes5*, *Hey1* and *Hey2* have more severe phenotypes than single mutants, suggesting additive effects between Hes/Hey repressors (Benito-Gonzalez and Doetzlhofer 2014; Li et al. 2008; Tateya et al. 2011; Zheng et al. 2000; Zine et al. 2001). Hence, the multiplicity of DSL ligands and Notch effectors makes the lateral inhibition of HC fate decisions a robust and relatively fail-safe mechanism, although some of its components (e.g. Dll1, Notch1) are clearly more critical than others to its operation.

Are Hair Cell Fate Decisions Biased? In Drosophila, a number of factors can provide a positive or negative bias in the signal-sending or signal-receiving abilities of cells interacting by lateral inhibition, but are these at play during HC fate decisions? The fact that there are strong differences in the expression levels of Atoh1 in prosensory cells (very low/absent) versus nascent HCs (very high) suggests that the latter might have a competitive advantage from the onset of lateral inhibition, but how this might be achieved is unknown. Numb, an endocytotic adaptor protein able to reduce the activity of the Notch receptor during asymmetric fate decisions in *Drosophila* (Couturier et al. 2013), does not appear to have such effect in the ear since its overexpression does not bias HC versus SC fate choices (Eddison et al. 2000; Eddison et al. 2015). The E3-ubiquitin ligase Mib, which is required for the internalization and activity of DSL ligands (Itoh et al. 2003), could in theory provide an advantage to some signal-sending cells, but this has not been directly tested.

Lateral Induction and Prosensory Specification

At the time the role of Jag2/Dll1 in the lateral inhibition of HC formation was uncovered, it became clear that an additional DSL ligand, Jag1, had a very distinct function. In fact, Jag1 is expressed long before HC differentiation by the prosensory cells and later by the SCs, which contact one another – a first hint that its expression is not repressed by Notch activity (Adam et al. 1998; Eddison et al. 2000; Morrison et al. 1999). Furthermore, the *Jag1* mutants had a very distinct phenotype from the *Dll1* or *Jag2*-deficient mice: they exhibited a circling behaviour, due to the absence of their vestibular cristae, and an organ of Corti with more inner HCs but fewer outer HCs than normal (Kiernan et al. 2001; Tsai et al. 2001). This, along with experiments showing that early and transient overexpression of an active form of Notch can induce the formation of ectopic sensory territories and Jag1 expression in the chick inner ear (Daudet and Lewis 2005), suggested that Jag1/Notch signalling regulates, by lateral induction, prosensory specification (Fig. 8.6).

Notch Activity Promotes Prosensory Specification by Maintaining Sox2 Expression Prosensory specification is the series of events leading to the formation of the prosensory cells, the population of otic progenitors competent to differentiate into HCs and SCs. Fate map experiments relying on the classic chick-quail grafting technique have shown that sensory organ progenitors are originally located within a large medial domain extending along the anteroposterior axis of the otic placode (Sánchez-Guardado et al. 2014). In the early otic vesicle, they are contained within a large 'pan-sensory' domain extending along the anteroposterior axis, which encompasses the anterior neurogenic patch and expresses among other markers

Prosensory specification

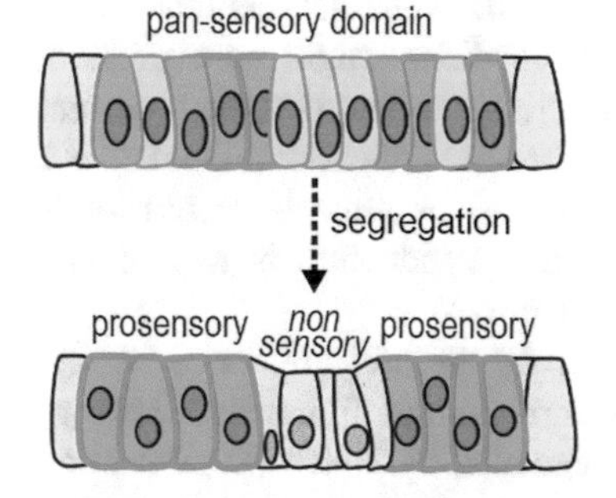

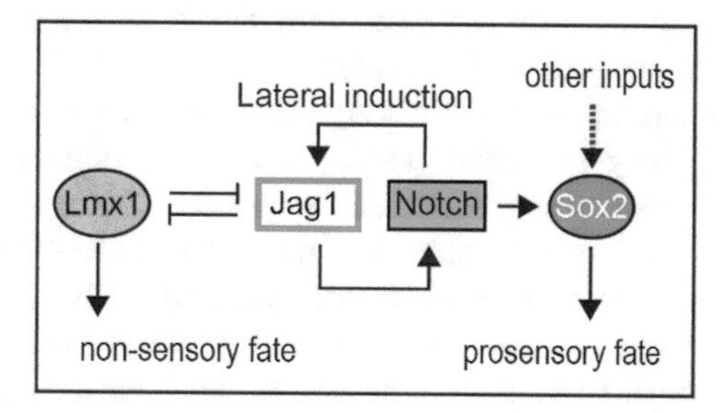

Fig. 8.6 Lateral induction during prosensory specification and a hypothetical model for sensory organ segregation. Left panel: Jag1 and Sox2 are initially expressed in a continuous 'pan-sensory' domain, but their expression is downregulated in prospective non-sensory territories during sensory organ segregation. The upregulation of Lmx1a could contribute to a reduction of Notch activity and Sox2 expression during the separation of sensory organs. Right panel: hypothetic regulatory circuit linking Notch activity, Lmx1 and Sox2 expression. The arrows do not imply direct regulation or interaction

FGF10, *Sox2* and *Jag1* (Adam et al. 1998; Alsina et al. 2004; Cole et al. 2000; Mann et al. 2017; Sánchez-Guardado et al. 2013). Next, distinct prosensory patches arise progressively by segregation from the pan-sensory domain – a mechanism that could have contributed to the multiplication and diversification of the inner ear sensory organs in the course of vertebrate evolution (Fritzsch and Beisel 2001). Thus, the original pool of sensory-competent cells must be maintained and expanded throughout formation of the sensory organs. Lateral induction is in theory ideally suited to fulfil these roles: the positive feedback loop linking Notch and Jag1 expression could (i) maintain Notch activity and prosensory character within interacting cells and (ii) specify new prosensory cells by recruiting 'naïve' cells (Notch-OFF) into a Notch-ON state. By large, the experimental evidence supports this idea but indicates that Notch activity is not sufficient for prosensory specification. In fact, overexpression of NICD in the early chick (Daudet and Lewis 2005; Neves et al. 2011) or mouse (Hartman et al. 2010; Liu et al. 2012a; Pan et al. 2010; Pan et al. 2013) otocyst does not convert all transfected cells into prosensory cells, and the capacity of Notch to induce ectopic sensory organs is restricted to a narrow developmental window (Liu et al. 2012a). Conversely, blocking Notch activity with GSI in cultures of chick otocysts strongly impairs, but does not completely prevent, sensory organ formation (Daudet et al. 2007). Likewise, the *RBPJ(k)/CSL* cKO mouse exhibits a severe atrophy of the vestibular organs but a partial formation of the cochlear prosensory domain (Basch et al. 2011; Yamamoto et al. 2011). Some studies, however, have reported that GSI or TACE inhibitor treatments can suppress prosensory specification in organotypic cultures of embryonic (E12-E13) organ of Corti (Hayashi et al. 2008; Munnamalai et al. 2012). These seemingly contradictory results may be explained by the differences in experimental approaches and the fact that a CSL-null may not be equivalent to a Notch-OFF condition, due to the requirement for CSL for the constitutive repression of some of the Notch target genes (Barolo et al. 2002): in the absence of CSL, the Notch targets that would normally be repressed by CSL in a Notch-OFF condition might be expressed, mimicking in fact a Notch-ON situation. Nevertheless, the fact that the recurrent consequence of blocking Jag1/Notch activity at early stages of ear development is a strong reduction (as opposed to a complete absence) of prosensory territories strongly suggests that the primary function of lateral induction is to maintain prosensory specification, rather than inducing it 'de novo'.

The prosensory factors targeted by Notch activity are still unknown, but are likely to include Sox2, which is required for the formation of all inner ear sensory organs (Kiernan et al. 2005) and is positively regulated by Notch (Daudet et al. 2007; Pan et al. 2010). The expression of Sox2 is widespread in the early otocyst before becoming restricted to the anterior neurosensory competent domain and the prosensory patches (Neves et al. 2007; Neves et al. 2011; Steevens et al. 2017). Furthermore, recent lineage-tracing studies relying on an inducible Sox2-CreER mouse model have confirmed that the early Sox2-expressing cells include prospective prosensory cells but also a large proportion of cells that will eventually lose Sox2 and give rise to non-sensory territories (Gu et al. 2016, 2; Steevens et al. 2019).

This reduction of Sox2 expression may be due to the localized dampening of lateral induction. In fact, Jag1 is downregulated in between segregating sensory organs in the chick and zebrafish otocyst (Ma and Zhang 2015; Mann et al. 2017), and if Notch activity is artificially increased, the cristae (as well as other sensory organs) fail to segregate (Mann et al. 2017). Thus, the spatial regulation of lateral induction could determine the number and size of sensory organs that segregate from the initial pan-sensory domain. This might explain why forcing Notch activity tends to 'induce' ectopic sensory territories close to the endogenous sensory organs (see, e.g. Hartman et al. 2010; Liu et al. 2012a): in these experiments, Notch overactivation could maintain prosensory character (and Sox2) in sensory-competent regions that would otherwise lose this character over time.

Besides Sox2 and Notch, many transcription factors and signalling pathways are likely to promote or antagonize prosensory specification (Żak et al. 2015). For example, the LIM-homeodomain transcription factor Lmx1a, expressed in non-sensory domains, antagonizes Notch activity. In the *Lmx1a*-null mouse, cells normally contributing to non-sensory territories in between sensory organs adopt (or retain) a prosensory character (Mann et al. 2017), producing fused sensory organs (Koo et al. 2009; Nichols et al. 2008). Several mutant mice exhibit defects in sensory organ segregation or size, and some of the underlying genes could impact on prosensory specification. The interactions between Notch signalling and Sox2 are most likely the tip of the iceberg, and much remains to be learnt about the genetic and epigenetic factors regulating prosensory specification.

Which Notch Components Mediate Its Prosensory Function? Jag1 is positively regulated by Notch activity and can activate its own expression 'in trans', which indicate that it functions by lateral induction (Daudet and Lewis 2005; Eddison et al. 2000; Hartman et al. 2010; Neves et al. 2011; Pan et al. 2010). However, this does not mean that it is the only DSL ligand capable of regulating prosensory specification. In fact, Jag2 (Neves et al. 2011) or Dll1 (Mann et al. 2017) can induce ectopic sensory patches when overexpressed at early stages of chick inner ear development. Furthermore, the inner ear of *Dll1*-null mice exhibits a very small (or absent) saccule (Brooker 2006; Kiernan 2005), whilst it is the only sensory patch that seems to develop normally in the *Jag1*-cKO mice (Brooker 2006; Kiernan et al. 2006). This suggests that Notch activity elicited by Dll1 during neurogenesis could contribute to the maintenance of the saccule progenitors. If several DSL can mediate the prosensory function of Notch, what about Notch receptors and their effectors? Receptors other than Notch1 must at least contribute to prosensory specification, since HCs do form (and in excess) in the *Notch1* cKO mouse. The identity of the Notch effectors is equally elusive. It was originally proposed that lateral induction may be mediated through Hey1 and Hey2 (also known as Hesr1/2), since they are expressed within the prosensory domain before HC formation in the organ of Corti (Hayashi et al. 2008) and the vestibular organs (Petrovic et al. 2015; Tateya et al. 2011). However, *Hey1/Hey2* double KO mice do not have any defect in prosensory specification or any reduction in HC numbers (Benito-Gonzalez and Doetzlhofer 2014). Prosensory specification is also unaffected in the triple KO for *Hes1*, *Hes5* and *Hey1* (Tateya et al. 2011). Some Notch receptor(s) and effector(s) specifically involved in lateral induction may eventually be identified. However, the great level of functional redundancy seen during lateral inhibition is a strong hint that Notch could promote prosensory specification through multiple ligands, receptors and effectors. It is also clear that some canonical Notch effectors are not solely regulated by Notch in the ear. For example, Hey1/2 expression are maintained in the cochlea of *RBPJK* cKO mice (Basch et al. 2011) and only partly reduced after pharmacological inhibition of Notch activity (Petrovic et al. 2015), whilst FGF, Hedgehog, BMP and Wnt signalling can also regulate their expression (Benito-Gonzalez and Doetzlhofer 2014; Doetzlhofer et al. 2009; Munnamalai et al. 2012; Petrovic et al. 2015; Tateya et al. 2013). Although we do not know if these transcriptional effects are direct or not, a number of signalling pathways seem able to 'hijack' traditional Notch effectors to impact on HC fate decisions as well as prosensory specification.

Notch Signalling During Otic Placode Formation

Notch1, Dll1 and Jag1 and some Notch effectors (Hes1, cHes5.2) are expressed in the pre-otic ectoderm on both sides of the hindbrain before the placode itself is morphologically recognizable (Jayasena et al. 2008; Jayasena et al. 2008; Myat et al. 1996; Shida et al. 2015). At this stage, Wnt and FGF signals secreted by the neighbouring tissues (neural tube, mesoderm, endoderm) promote otic induction (reviewed in Chen and Streit 2013), and Pax2-positive precursors for the otic and epibranchial placodes are still intermingled in the pre-otic territory (Groves and Bronner-Fraser 2000; Streit 2002). Although it is tempting to imagine that Notch signalling could regulate the specification of these different cell types, experimental manipulation of Notch activity produced conflicting results. In the mouse, sustained NICD overexpression in the Pax2 lineage leads to the expansion of an otic placode-like epithelium, which expresses the otic marker Pax8, but not others such as Pax2, Sox9,Gbx2 and Hmx3 (Jayasena et al. 2008). Since the placode is slightly reduced in size in a partial *Notch1* mutant, the authors proposed that Notch activity could promote otic placode formation, possibly by augmenting Wnt signalling (Jayasena et al. 2008). However, in chick embryos, overexpression of NICD inhibits otic placode formation, whilst cells in which Notch is reduced (by expressing a dominant-negative form of Dll1 or CSL) tend to adopt an otic fate (Shida et al. 2015). These differences may be due to species-specific functions for Notch or to the distinct experimental approaches used, since electroporation in the chick embryo would affect more cells than those of the Pax2 lineage. Another uncertainty is the mode of action (inhibition, induction or something else) of Notch in this context, since the consequences of blocking Notch activity on the expression of Dll1 and Jag1 have not been investigated. More work is therefore necessary to clarify the roles of Notch signalling during otic placode formation.

Managing and Multitasking: Specificity and Integration of Notch Functions in the Developing Inner Ear

The diversity of functions fulfilled by Notch signalling in the inner ear is remarkable, and more often than not, these overlap in space and time. How then are the effects of different ligands and various modes of signalling integrated and converted into distinct responses?

Competence States Determine the Context-Specific Effects of Notch Signalling

What determines the specific cell identities adopted through lateral inhibition or induction is the intrinsic competence of the interacting cells, which can be defined as their capacity to adopt a particular fate (or express a set of transcription factors) when Notch is ON or OFF. The competence of the neurosensory progenitors changes during development. In mammals and birds, they produce neuroblasts first and then HC and SC. This transition is at least partly dependent on antagonistic cross-interactions between Neurog1 and Atoh1: in the *Neurog1*-null mouse, some of the prospective neuroblasts adopt a prosensory character and upregulate Atoh1 (Raft et al. 2007), and HCs form earlier than expected (Matei et al. 2005). Among the other potential regulators of Atoh1 and Neurog1 are the inhibitors of differentiation (Id1–4) HLH factors, which act as dominant-negative regulators of the proneural bHLH factors (reviewed in Jones 2004). All Ids are present in the developing inner ear (Jones 2006; Kamaid et al. 2010; Ozeki et al. 2007), and the overexpression of Id1–3 in ovo (Kamaid et al. 2010) or in organotypic cultures of mouse organ of Corti (Jones et al. 2006) can prevent HC differentiation, presumably by blocking Atoh1 activity. However, the absence of *Id1* and *Id3* in double KO mice does not produce obvious defects in prosensory specification or HC differentiation, at least in the cochlea (Jones 2006), suggesting an important level of redundancy between *Id* genes. Another factor that probably regulates the competence state of neurosensory progenitors is Sox2: it is required for their formation and can directly regulate Atoh1 expression, but it can also antagonize its ‘pro-HC’ and ‘proneural’ effects when expressed at high levels (reviewed in Gálvez et al. 2017; Raft and Groves 2014), which suggests dose-dependent effects very similar to those described during central neurogenesis (Pevny and Nicolis 2010).

The Epigenetic Landscape: A Regulator of Neurosensory Competence? Interactions between transcription factors are critical for setting up a competence state, but their ability to access their genomic targets is dependent on chromatin structure. The competence of otic progenitors is therefore expected to be sensitive to epigenetic modifications such as post-translational modifications of histones or DNA methylation. Explorations of this complex level of regulation have started relatively recently, but there are already some indications of its importance (Doetzlhofer and Avraham 2017). For example, de novo mutations in the *Chd7* gene, which encodes a chromatin remodelling enzyme, cause the CHARGE syndrome, associated with congenital defects in inner ear morphology and function. During otic development, CHD7 expression becomes progressively restricted to the neurosensory regions (Hurd et al. 2010). The phenotype of *Chd7*-deficient mice is dose-dependent and complex, but it includes defects in neurogenesis and vestibular morphogenesis (absence of cristae and semi-circular canals) as well as a reduction of Sox2 and Jag1 expression (Hurd et al. 2012), consistent with abnormal neurosensory specification. Are there any epigenetic modifications specifically associated with the specification of the neurosensory lineage? A recent study compared the open chromatin landscape of FACS-sorted Sox2-positive and negative cells in the embryonic mouse cochlea and found that approximately 29,300 ATAC-Seq peaks were enriched in the prosensory cell population (Wilkerson et al. 2019), suggesting that may be the case. Further studies on more restricted populations of cells, or even single cells, will be needed to confirm these first insights and to determine if and how epigenetic marks influence the competence states of otic progenitors.

In summary, the competence state of otic neurosensory progenitors is the product of the interactions between bHLH proneural genes, Sox2, their regulators and the epigenetic factors impinging on their activities in a context-specific manner. What a cell does in response to Notch activity will change according to its competence state, which explains why blocking or activating Notch activity produces different outcomes at different stages of inner ear development.

Integration of Notch Activity from Multiple Ligands and Modes of Signalling

Notch Effectors Are Activated in a Dose-Dependent Manner

There are several DSL ligands expressed in the inner ear, regulated in opposite ways, which raises the question of how their effects are integrated within signal-receiving cells. We do not know if there are any receptor(s) other than Notch1 mediating lateral inhibition and induction in the ear. Nevertheless, we can infer from other systems that even a single type of Notch receptor can be differentially activated by distinct DSL ligands. For example, biochemical assays have shown that Dll4 can bind 10 times more strongly than Dll1 to the Notch1 receptor (Andrawes et al. 2013); in cell lines carrying fluorescent reporters of Notch activity, Dll1 and Dll4 trigger, respectively, transient versus sustained activation of the Notch1 receptor (Nandagopal et al. 2018). In the inner ear, Jag1 and Dll1 may also bind differently to Notch1. In fact, the amount of cleaved Notch1ICD increases drastically within sensory patches at the time of HC formation (Murata et al. 2006), suggesting that Jag1 is a relatively weak ligand for Notch1 compared to Dll1 or Jag2. Another observation that fits with this idea is that whilst Dll1 immunostaining is detected exclusively within intracellular vesicles, Jag1 tends to accumulate at the cell membrane of prosensory cells (Chrysostomou et al. 2012). This suggests that the internalization of Dll1 and therefore its trans-signalling effect is stronger than that of Jag1. The strength of Notch activation could next be translated into a dose-dependent expression of Hes/Hey effectors: in chick otocysts, the expression of *Hes5* is reduced faster than that of *Hey1* following GSI treatment, implying that *Hes5* requires higher levels of Notch activity for its induction (Petrovic et al. 2015). Comparable findings were obtained in the mouse organ of Corti: the pharmacological inhibition of Notch activity induces dose-dependent changes in the expression of its *Hes/Hey* effectors, and *Hey* genes are less sensitive than *Hes* genes to this blockade (Basch et al. 2016b; Doetzlhofer et al. 2009; Maass et al. 2015). It is therefore possible that inputs from different Notch receptors are simply added and converted into a dose-dependent combination of *Hes/Hey* genes. The subtype of NICD produced may not be critical, since substituting the Notch1ICD by the Notch2ICD in transgenic mice (on a mixed genetic background) leads to the formation of a normal-looking organ of Corti, although the hybrid receptor is not as efficient as the intact Notch1 receptor (Liu et al. 2015).

Ligand-Receptor Interactions Are Modulated by Fringe-Dependent Glycosylation If a common set of transcriptional targets is regulated by several Notch receptors in a dose-dependent manner, it is at the DSL/Notch binding and activation steps that the "specific" effects of a given ligand/receptor pair are determined: the more the ligand-receptor interactions, the stronger the Notch activation, and the more the *Hes/Hey* genes are induced. Besides the amount of cell-surface ligand and receptor, factors modifying these interactions could potentially 'tune' a given Notch receptor to a specific ligand. Among these are Fringe proteins, which are expressed in the developing ear (Cole et al. 2000; Morsli et al. 1998). The initial analysis of a *Lfng* mutant mouse cochlea showed no HC patterning defects, but surprisingly, the absence of *Lfng* could partly rescue the overproduction of inner HCs elicited by *Jag2* absence (Zhang et al. 2000). The authors proposed that Lfng could reduce Dll1/Notch activation and its absence could perhaps restore normal levels of lateral inhibition in the absence of Jag2, although other scenarios are possible. In a recent study, Basch et al. (2016b) revisited the roles of Lfng and its paralogue *Manic Fringe* (*Mfng*) during cochlear development. They found that Lfng is dynamically expressed from the medial to the lateral side of the prosensory domain during its specification and then becomes enriched in SCs; Mfng is expressed later and is restricted to HCs. However, there is transient co-expression of Lfng and Mfng at the medial boundary of the organ of Corti, and when both genes are inactivated, supernumerary inner HCs (and their SCs) are produced, resembling the *Jag2* mutant phenotype and other mutants with a partial loss of Notch function in the ear (Basch et al. 2016b), including the *Jag1* cKO (Brooker 2006; Kiernan et al. 2006). This suggests that Mfng and Lfng increase Notch1 sensitivity to its ligands and are required for robust lateral inhibition in the inner HCs region, but further work will be needed to establish which particular ligand-receptor interactions are modulated by Lfng/Mfng. An interesting hypothesis raised by the authors is that Fringe could modulate cis-inhibition, which is the capacity of DSL ligands to inhibit the activation of Notch receptors 'in cis'. We do not know yet if there is any ligand acting in this way in the ear, but Dll1 seems to be a relatively poor cis-inhibitor, at least in the chick inner ear: its overexpression under the control of either a Notch-responsive or constitutive promoter can produce large patches of transfected cells without HCs (Chrysostomou et al. 2012). This result suggests that Dll1 elicits Notch activity *in trans* but does not prevent Notch activation *in cis* (otherwise more/all Dll1 expressing cells would be HCs), although further work is needed to validate this conclusion.

Making Sense of Notch Signalling in the Ear: Mathematical Models to the Rescue As we progress in our description of the mechanisms of Notch signalling in the ear, we are faced with an increasingly complex and dynamic picture: multiple ligands regulated in a different way, acting *in trans* or *in cis*, activating one or more Notch receptors, regulating the expression of several *Hes/Hey* and proneural genes, themselves subjected to regulation by external factors and feeding back on ligand expression! Evidently, changing the activity of one component can generate phenotypes that are difficult to interpret. To understand the logic of Notch signalling at a system level, we need mathematical models. The first model of lateral inhibition by Collier et al. (1996) used differential equations to compute ligand expression as a function of Notch activation. Their simulations showed that starting from random conditions, an alternated mosaic of (computer) cells with either high-Notch/low-ligand or low-Notch/high-ligand could be formed if there was a strong enough negative feedback loop between Notch activity and ligand expression in signal-receiving cells. In an extension of this model, Petrovic et al. (2014) studied the interplay between lateral inhibition and induction by introducing in the model two ligands, regulated in opposite manner. They found that the two modes of signalling can co-exist and produce a cellular mosaic only if (i) lateral induction does not elicit too high levels of Notch activity and (ii) the intercellular lateral inhibition feedback loop is reinforced by the positive autoregulation of the proneural gene that it controls. These predictions agree with the differential expression of Hes/Hey genes in response to Dll1 versus Jag1 in the ear (Petrovic et al. 2014; Petrovic et al. 2015) and suggest that the Atoh1 autoregulatory loop (Abdolazimi et al. 2016; Bermingham et al. 1999; Helms et al. 2000; Woods et al. 2004) is critical for robust lateral inhibition. Another recent study investigated the impact of apical surface area on the outcomes of lateral inhibition (Shaya et al. 2017). Using in silico simulations and experiments conducted in cell lines, the authors proposed that smaller cells were more likely to act as signal-sending cells during lateral inhibition. One observation supporting this conclusion was that immature HCs have smaller apical surfaces compared to that of SCs in the embryonic basilar papilla, although we do not know whether this is also the case for uncommitted HCs. Models are powerful discovery tools, and their predictions, at times counter-intuitive, can open up new lines of enquiries: Would manipulation of cell size be sufficient to bias HC fate decisions? What would happen if the auto-regulatory feedback loop controlling Atoh1 expression was altered or if the strength of lateral induction or inhibition was artificially boosted? With the recent progress in genome editing technologies, these questions can now be tested experimentally in a wide range of model organisms.

Integration of Notch Signalling with Cell Proliferation and Rearrangements

Developing tissues change considerably in form through cell addition and changes in cell morphology and position. These can in turn impact on the intercellular signalling pathways dependent on diffusible or cell-surface molecules. In the mammalian cochlea, convergent-extension movements responsible for the elongation of the epithelium cause local cell rearrangements during cell differentiation (Yamamoto et al. 2009). The situation is even more dynamic in the vestibular organs, where proliferation and HC production occur simultaneously and for an extended period of time (Burns et al. 2012; Goodyear et al. 1999). Thus, lateral inhibition operates in a changing cellular environment, which could explain why transient contacts between immature and more mature HCs are frequently observed in developing sensory epithelia (Chrysostomou et al. 2012; Goodyear and Richardson 1997). How then are HCs ultimately positioned to their right place? One solution, originally supported by computational modelling (Podgorski et al. 2007), could be some form of differential adhesion that would result in HCs repelling one another whilst adhering more strongly to SC. Strong candidates to mediate this function belong to the nectin family of immunoglobulin-like adhesion molecules. In the organ of Corti, nectin1 is expressed in HCs and binds strongly to nectin3, present in SCs; their inactivation in KO mice results in patterning errors, in particular in the outer HC region, but without significant changes in the number of HCs or SCs produced (Fukuda et al. 2014; Togashi et al. 2011).

Notch signalling may conversely, through some of its transcriptional targets, impact on cell proliferation. The expression of p27Kip1, which triggers the cell cycle exit of prosensory cells, is, for example, downregulated in the *Jag1* cKO, in which IHCs and their SCs are produced in excess (Brooker 2006). Likewise, the absence of *Hey/Hes* genes can result in increased cell proliferation in the organ of Corti (Tateya et al. 2011), although the mechanistic link between Notch and the cell cycle has not been established.

Notch Signalling and Hair Cell Regeneration

Hair cells are vulnerable to ototoxic drugs and acoustic trauma, and they spontaneously die during ageing. In mammals, the vestibular organs have a limited capacity for HC replacement, but the auditory HCs (approximately 15,000 in humans) are only produced during development. Their disappearance leads to irreversible hearing loss, for which the only available treatments are hearing aids or cochlear implants (reviewed in Géléoc and Holt 2014). In contrast, non-mammalian vertebrates can regenerate HCs after damage throughout life. In the avian basilar papilla, a classic model for regeneration studies, SCs are normally quiescent, but these are in fact 'tissue stem cells' that can after tissue damage (i) transdifferentiate directly into HCs and (ii) re-enter the cell cycle to produce new HCs and new SCs (reviewed in Rubel et al. 2013; Stone and Cotanche 2007). Notch signalling is reactivated in the damaged basilar papilla during HC regeneration: Atoh1, DSL ligands and Notch effectors are upregulated 1 day after HC loss (Cafaro et al. 2007; Daudet et al. 2009; Stone and Rubel 1999). Furthermore, blocking Notch activity with GSI during regeneration causes an upregulation of Atoh1 and excess transdifferentiation of SCs into HCs in vitro (Daudet et al. 2009; Lewis et al. 2012). This suggests that lateral inhibition regulates the regeneration of HCs in a very similar manner to what it does during their embryonic formation. But importantly, GSI do not induce spontaneous regeneration in an undamaged epithelium, which implies that Notch activity does not maintain SCs quiescent (Daudet et al. 2009). The situation differs in the chicken utricle, where the continuous turnover of HCs is associated with mosaic Dll1 expression (Stone and Rubel 1999). There, GSI or ADAM-10 metalloprotease inhibitors increase HC regeneration after damage but also cause SC proliferation in the intact tissue (Warchol et al. 2017). Interestingly, GSI also stimulate SC proliferation after HC damage within the neuromasts of the zebrafish lateral line, another model for HC regeneration studies (Ma et al. 2008). However, we do not know for certain if these effects are due to inhibition of Notch activity: GSI and ADAM-10 metalloprotease inhibitors can interfere with the processing of other transmembrane molecules that could potentially impact on cell proliferation. Further studies relying on genetic manipulation of Notch activity will be required to clarify this point.

The Limited Regenerative Potential of Mammalian Vestibular Organs Is Improved by Notch Inhibition The adult mammalian vestibular system has a limited capacity for spontaneous HC regeneration (Forge et al. 1993; Warchol et al. 1993), which is thought to rely on transdifferentiation mainly (Rubel et al. 2013). This was recently confirmed using a genetic method for ablating utricular HCs and tracing the SCs: there is a very slow addition of new HCs (about two per week) in the adult mouse utricle, and SCs can convert into new HCs after extensive tissue damage (Bucks et al. 2017). We do not know which Notch receptors are expressed in the adult vestibular organs, but the utricle and cristae express at least Jag1 (Oesterle et al. 2008; Wang et al. 2010) and *Hes1/5* (Hartman et al. 2009; Slowik and Bermingham-McDonogh 2013). Furthermore, Atoh1 is upregulated following ototoxic damage, and interfering with Notch activity can improve the regenerative response of mammalian vestibular organs in vivo and in vitro (Jung et al. 2013; Lin et al. 2011; Slowik and Bermingham-McDonogh 2013; Wang et al. 2010). However the capacity of SCs to convert to HCs in response to Notch inhibition remains very low in adult compared to neonate vestibular organs (Collado et al. 2011), including in organotypic cultures derived from human tissue (Taylor et al. 2018). Combining

Notch inhibition with overexpression of Atoh1 is not sufficient to induce robust HC regeneration either (Lin et al. 2011; Taylor et al. 2018), suggesting that other signals restrict the competence of vestibular SCs to transdifferentiate into HCs.

Notch Activity Is Turned Off in the Adult Organ of Corti: Implications for Regenerative Therapies Any report of successful auditory HC regeneration is invariably met, for good reasons, with a mix of excitement and scepticism. Where do we stand today with respect to the possibility to use Atoh1 or Notch-targeted therapies for hearing loss? The bulk of the evidence indicates that the capacity of Atoh1 to trigger ectopic HC differentiation (Zheng and Gao 2000) is limited to stages when SCs are still relatively immature (Kelly et al. 2012; Liu et al. 2012b); in the adult cochlea, there is no conclusive evidence that Atoh1 overexpression can lead to the regeneration of functional HCs (reviewed in Richardson and Atkinson 2015). Likewise, the initial report that intra-cochlear administration of GSI may stimulate HC regeneration in adult mice exposed to acoustic trauma (Mizutari et al. 2013) has not been replicated so far. A major caveat with regeneration studies is the variability of HC death induced by ototoxic drugs or acoustic trauma: if HCs are observed several weeks after damage, are these surviving HCs or regenerated ones? One way to answer this question is to analyse tissue at different times post-treatment. If regeneration occurs, one should find HCs with immature features (e.g. short stereociliary bundles; expression of Atoh1) soon after damage, just as in the vestibular organs. None of the studies claiming successful regeneration of auditory HCs (in the adult epithelium) has provided such evidence. Finally, the expression of Notch components decreases rapidly during the maturation of the organ of Corti, as does the capacity of SCs to convert into HCs in response to GSI treatment (Maass et al. 2015; Maass et al. 2016). Even if a very low level of Notch activity was present in the adult damaged cochlea, it may not necessarily regulate HC regeneration. If Notch has context-dependent transcriptional targets and functions in the course of ear development, why should it be different at adult stages? The uncertainties about the molecular targets and therapeutic effects of GSI in the adult organ of Corti are valid reasons to advocate caution with regard to the ongoing clinical trials in hearing loss sufferers. The recent boom in private and public investment for regenerative inner ear therapies (reviewed in Schilder et al. 2019) is welcome and justified by the major economical and societal impact of hearing loss, but developing a successful cure will depend on our capacity to recognize (and fill) our gaps in the basic knowledge of HC development and regeneration.

Conclusion

Notch signalling has diverse and intertwined functions in the development and regeneration of the inner ear. As is the case in other tissues, the context-specific outcomes of Notch signalling in the ear are determined by the mode of regulation and signalling abilities of the DSL ligands but also, and crucially, by the competence state of the signalling cells, which changes as cells progress along their differentiation route. Understanding the genetic and epigenetic mechanisms responsible for these transitions could improve the prospect of regenerative therapies for inner ear disorders. Besides exploring the molecular interactions between Notch and the other major signalling pathways (FGF, Wnt, Hedgehog, etc.), we need a deeper understanding of their integration with the cellular and physiological aspects of inner ear development. Whilst this entails significant intellectual and experimental challenges, the ongoing progress in single-cell and functional genomics, imaging, inner ear organoid cultures and computational modelling are opening up new and exciting venues for inner ear (and Notch) *aficionados*.

References

Abdolazimi Y, Stojanova Z, Segil N (2016) Selection of cell fate in the organ of Corti involves the integration of Hes/Hey signaling at the Atoh1 promoter. Development 143:841–850

Abelló G, Khatri S, Giráldez F, Alsina B (2007) Early regionalization of the otic placode and its regulation by the Notch signaling pathway. Mech Dev 124:631–645

Adam J, Myat A, Le Roux I, Eddison M, Henrique D, Ish-Horowicz D, Lewis J (1998) Cell fate choices and the expression of Notch, Delta and Serrate homologues in the chick inner ear: parallels with Drosophila sense-organ development. Development 125:4645–4654

Alsina B, Whitfield TT (2017) Sculpting the labyrinth: morphogenesis of the developing inner ear. Semin Cell Dev Biol 65:47

Alsina B, Abelló G, Ulloa E, Henrique D, Pujades C, Giraldez F (2004) FGF signaling is required for determination of otic neuroblasts in the chick embryo. Dev Biol 267:119–134

Ambler CA, Watt FM (2010) Adult epidermal Notch activity induces dermal accumulation of T cells and neural crest derivatives through upregulation of jagged 1. Development 137:3569–3579

Andrawes MB, Xu X, Liu H, Ficarro SB, Marto JA, Aster JC, Blacklow SC (2013) Intrinsic Selectivity of Notch 1 for Delta-like 4 Over Delta-like 1. J Biol Chem 288:25477–25489

Barolo S, Stone T, Bang AG, Posakony JW (2002) Default repression and Notch signaling: Hairless acts as an adaptor to recruit the corepressors Groucho and dCtBP to Suppressor of Hairless. Genes Dev 16:1964–1976

Basch ML, Ohyama T, Segil N, Groves AK (2011) Canonical Notch signaling is not necessary for prosensory induction in the mouse cochlea: insights from a conditional mutant of RBPj. J Neurosci 31:8046–8058

Basch ML, Brown RM, Jen H-I, Groves AK (2016a) Where hearing starts: the development of the mammalian cochlea. J Anat 228:233–254

Basch ML, Ii RMB, Jen H-I, Semerci F, Depreux F, Edlund RK, Zhang H, Norton CR, Gridley T, Cole SE et al (2016b) Fine-tuning of Notch signaling sets the boundary of the organ of Corti and establishes sensory cell fates. elife 5:e19921

Benito-Gonzalez A, Doetzlhofer A (2014) Hey1 and Hey2 control the spatial and temporal pattern of mammalian auditory hair cell differentiation downstream of hedgehog signaling. J Neurosci 34:12865–12876

Bermingham NA, Hassan BA, Price SD, Vollrath MA, Ben-Arie N, Eatock RA, Bellen HJ, Lysakowski A, Zoghbi HY (1999) Math1: an essential gene for the generation of inner ear hair cells. Science 284:1837–1841

Bray SJ (2006) Notch signalling: a simple pathway becomes complex. Nat Rev Mol Cell Biol 7:678–689

Brooker R (2006) Notch ligands with contrasting functions: Jagged1 and Delta1 in the mouse inner ear. Development 133:1277–1286

Bucks SA, Cox BC, Vlosich BA, Manning JP, Nguyen TB, Stone JS (2017) Supporting cells remove and replace sensory receptor hair cells in a balance organ of adult mice. elife 6:e18128
Burns JC, On D, Baker W, Collado MS, Corwin JT (2012) Over half the hair cells in the mouse utricle first appear after birth, with significant numbers originating from early postnatal mitotic production in peripheral and striolar growth zones. JARO 13:609–627
Cafaro J, Lee GS, Stone JS (2007) Atoh1 expression defines activated progenitors and differentiating hair cells during avian hair cell regeneration. Dev Dyn 236:156–170
Cai T, Seymour ML, Zhang H, Pereira FA, Groves AK (2013) Conditional deletion of Atoh1 reveals distinct critical periods for survival and function of hair cells in the organ of Corti. J Neurosci 33:10110–10122
Chen J, Streit A (2013) Induction of the inner ear: stepwise specification of otic fate from multipotent progenitors. Hear Res 297:3–12
Chen P, Johnson JE, Zoghbi HY, Segil N (2002) The role of Math1 in inner ear development: uncoupling the establishment of the sensory primordium from hair cell fate determination. Development 129:2495–2505
Chrysostomou E, Gale JE, Daudet N (2012) Delta-like 1 and lateral inhibition during hair cell formation in the chicken inner ear: evidence against cis-inhibition. Development 139:3764–3774
Cole LK, Le Roux I, Nunes F, Laufer E, Lewis J, Wu DK (2000) Sensory organ generation in the chicken inner ear: contributions of bone morphogenetic protein 4, serrate1, and lunatic fringe. J Comp Neurol 424:509–520
Collado MS, Thiede BR, Baker W, Askew C, Igbani LM, Corwin JT (2011) The postnatal accumulation of junctional E-cadherin is inversely correlated with the capacity for supporting cells to convert directly into sensory hair cells in mammalian balance organs. J Neurosci 31:11855–11866
Collier JR, Monk NA, Maini PK, Lewis JH (1996) Pattern formation by lateral inhibition with feedback: a mathematical model of delta-notch intercellular signalling. J Theor Biol 183:429–446
Cornell RA, Eisen JS (2005) Notch in the pathway: the roles of Notch signaling in neural crest development. Semin Cell Dev Biol 16:663–672
Couturier L, Mazouni K, Schweisguth F (2013) Inhibition of Notch recycling by numb: relevance and mechanism(s). Cell Cycle 12:1647–1648
Dahmann C, Oates AC, Brand M (2011) Boundary formation and maintenance in tissue development. Nat Rev Genet 12:43–55
Daudet N, Lewis JH (2005) Two contrasting roles for Notch activity in chick inner ear development: specification of prosensory patches and lateral inhibition of hair-cell differentiation. Development 132:541–551
Daudet N, Ariza-McNaughton L, Lewis J (2007) Notch signalling is needed to maintain, but not to initiate, the formation of prosensory patches in the chick inner ear. Development 134:2369–2378
Daudet N, Gibson R, Shang J, Bernard A, Lewis J, Stone J (2009) Notch regulation of progenitor cell behavior in quiescent and regenerating auditory epithelium of mature birds. Dev Biol 326:86–100
de Celis JF, Bray S (1997) Feed-back mechanisms affecting Notch activation at the dorsoventral boundary in the Drosophila wing. Development 124:3241–3251
de Celis JF, Garcia-Bellido A, Bray SJ (1996) Activation and function of Notch at the dorsal-ventral boundary of the wing imaginal disc. Development 122:359–369
del Álamo D, Rouault H, Schweisguth F (2011) Mechanism and significance of cis-inhibition in Notch signalling. Curr Biol 21:R40–R47
Doetzlhofer A, Avraham KB (2017) Insights into inner ear-specific gene regulation: epigenetics and non-coding RNAs in inner ear development and regeneration. Semin Cell Dev Biol 65:69–79
Doetzlhofer A, Basch ML, Ohyama T, Gessler M, Groves AK, Segil N (2009) Hey2 regulation by FGF provides a notch-independent mechanism for maintaining pillar cell fate in the organ of Corti. Dev Cell 16:58–69
Driver EC, Sillers L, Coate TM, Rose MF, Kelley MW (2013) The Atoh1-lineage gives rise to hair cells and supporting cells within the mammalian cochlea. Dev Biol 376:86–98

Eddison M, Le Roux I, Lewis J (2000) Notch signaling in the development of the inner ear: lessons from Drosophila. Proc Natl Acad Sci 97:11692–11699

Eddison M, Weber SJ, Ariza-McNaughton L, Lewis J, Daudet N (2015) Numb is not a critical regulator of Notch-mediated cell fate decisions in the developing chick inner ear. Front Cell Neurosci 9:74

Forge A, Li L, Corwin JT, Nevill G (1993) Ultrastructural evidence for hair cell regeneration in the mammalian inner ear. Science 259:1616–1619

Fritzsch B, Beisel KW (2001) Evolution and development of the vertebrate ear. Brain Res Bull 55:711–721

Fukuda T, Kominami K, Wang S, Togashi H, Hirata K -i, Mizoguchi A, Rikitake Y, Takai Y (2014) Aberrant cochlear hair cell attachments caused by Nectin-3 deficiency result in hair bundle abnormalities. Development 141:399–409

Gálvez H, Abelló G, Giraldez F (2017) Signaling and transcription factors during inner ear development: the generation of hair cells and Otic neurons. Front Cell Dev Biol 5

Géléoc GSG, Holt JR (2014) Sound strategies for hearing restoration. Science 344:1241062

Goodyear R, Richardson G (1997) Pattern formation in the basilar papilla: evidence for cell rearrangement. J Neurosci 17:6289–6301

Goodyear RJ, Gates R, Lukashkin AN, Richardson GP (1999) Hair-cell numbers continue to increase in the utricular macula of the early posthatch chick. J Neurocytol 28:851–861

Groves AK, Bronner-Fraser M (2000) Competence, specification and commitment in otic placode induction. Development 127:3489–3499

Gu R, Brown RM II, Hsu C-W, Cai T, Crowder AL, Piazza VG, Vadakkan TJ, Dickinson ME, Groves AK (2016) Lineage tracing of Sox2-expressing progenitor cells in the mouse inner ear reveals a broad contribution to non-sensory tissues and insights into the origin of the organ of Corti. Dev Biol 414:72–84

Haddon C, Jiang Y-J, Smithers L, Lewis J (1998) Delta-Notch signalling and the patterning of sensory cell differentiation in the zebrafish ear: evidence from the mind bomb mutant. Development 125:4637–4644

Hao J, Koesters R, Bouchard M, Gridley T, Pfannenstiel S, Plinkert PK, Zhang L, Praetorius M (2012) Jagged1-mediated Notch signaling regulates mammalian inner ear development independent of lateral inhibition. Acta Otolaryngol 132:1028–1035

Hartman BH, Hayashi T, Nelson BR, Bermingham-McDonogh O, Reh TA (2007) Dll3 is expressed in developing hair cells in the mammalian cochlea. Dev Dyn 236:2875–2883

Hartman BH, Basak O, Nelson BR, Taylor V, Bermingham-McDonogh O, Reh TA (2009) Hes5 expression in the postnatal and adult mouse inner ear and the drug-damaged cochlea. JARO 10:321–340

Hartman BH, Reh TA, Bermingham-McDonogh O (2010) Notch signaling specifies prosensory domains via lateral induction in the developing mammalian inner ear. Proc Natl Acad Sci 107:15792–15797

Hayashi T, Kokubo H, Hartman BH, Ray CA, Reh TA, Bermingham-McDonogh O (2008) Hesr1 and Hesr2 may act as early effectors of Notch signaling in the developing cochlea. Dev Biol 316:87–99

Helms AW, Abney AL, Ben-Arie N, Zoghbi HY, Johnson JE (2000) Autoregulation and multiple enhancers control Math1 expression in the developing nervous system. Development 127:1185–1196

Henrique D, Schweisguth F (2019) Mechanisms of Notch signaling: a simple logic deployed in time and space. Development 146:dev172148

Hurd EA, Poucher HK, Cheng K, Raphael Y, Martin DM (2010) The ATP-dependent chromatin remodeling enzyme CHD7 regulates pro-neural gene expression and neurogenesis in the inner ear. Development 137:3139–3150

Hurd EA, Micucci JA, Reamer EN, Martin DM (2012) Delayed fusion and altered gene expression contribute to semicircular canal defects in Chd7 deficient mice. Mech Dev 129:308–323

Itoh M, Kim C-H, Palardy G, Oda T, Jiang Y-J, Maust D, Yeo S-Y, Lorick K, Wright GJ, Ariza-McNaughton L et al (2003) Mind bomb is a ubiquitin ligase that is essential for efficient activation of Notch signaling by Delta. Dev Cell 4:67–82

Jarman AP, Groves AK (2013) The role of Atonal transcription factors in the development of mechanosensitive cells. Semin Cell Dev Biol 24:438–447
Jayasena CS, Ohyama T, Segil N, Groves AK (2008) Notch signaling augments the canonical Wnt pathway to specify the size of the otic placode. Development 135:2251–2261
Jones S (2004) An overview of the basic helix-loop-helix proteins. Genome Biol 5:226–226
Jones JM (2006) Inhibitors of differentiation and DNA binding (Ids) regulate Math1 and hair cell formation during the development of the organ of Corti. J Neurosci 26:550–558
Jones JM, Montcouquiol M, Dabdoub A, Woods C, Kelley MW (2006) Inhibitors of differentiation and DNA binding (Ids) regulate Math1 and hair cell formation during the development of the organ of Corti. J Neurosci 26:550–558
Jung JY, Avenarius MR, Adamsky S, Alpert E, Feinstein E, Raphael Y (2013) siRNA targeting Hes5 augments hair cell regeneration in aminoglycoside-damaged mouse utricle. Mol Ther 21:834–841
Kamaid A, Neves J, Giraldez F (2010) Id gene regulation and function in the prosensory domains of the chicken inner ear: a link between bmp signaling and Atoh1. J Neurosci 30:11426–11434
Kelly MC, Chang Q, Pan A, Lin X, Chen P (2012) Atoh1 directs the formation of sensory mosaics and induces cell proliferation in the postnatal mammalian cochlea in vivo. J Neurosci 32:6699–6710
Kiernan AE (2005) The Notch ligands DLL1 and JAG2 act synergistically to regulate hair cell development in the mammalian inner ear. Development 132:4353–4362
Kiernan AE, Ahituv N, Fuchs H, Balling R, Avraham KB, Steel KP, de Angelis MH (2001) The Notch ligand Jagged1 is required for inner ear sensory development. Proc Natl Acad Sci 98:3873–3878
Kiernan AE, Pelling AL, Leung KKH, Tang ASP, Bell DM, Tease C, Lovell-Badge R, Steel KP, Cheah KSE (2005) Sox2 is required for sensory organ development in the mammalian inner ear. Nature 434:1031–1035
Kiernan AE, Xu J, Gridley T (2006) The Notch ligand JAG1 is required for sensory progenitor development in the mammalian inner ear. PLoS Genet 2:e4
Kim WY, Fritzsch B, Serls A, Bakel LA, Huang EJ, Reichardt LF, Barth DS, Lee JE (2001) NeuroD-null mice are deaf due to a severe loss of the inner ear sensory neurons during development. Development 128:417–426
Koo SK, Hill JK, Hwang CH, Lin ZS, Millen KJ, Wu DK (2009) Lmx1a maintains proper neurogenic, sensory, and non-sensory domains in the mammalian inner ear. Dev Biol 333:14–25
Koundakjian EJ, Appler JL, Goodrich LV (2007) Auditory neurons make stereotyped wiring decisions before maturation of their targets. J Neurosci 27:14078–14088
Lanford PJ, Lan Y, Jiang R, Lindsell C, Weinmaster G, Gridley T, Kelley MW (1999) Notch signalling pathway mediates hair cell development in mammalian cochlea. Nat Genet 21:289–292
Lanford PJ, Shailam R, Norton CR, Ridley T, Kelley MW (2000) Expression of Math1 and HES5 in the cochleae of wildtype and Jag2 mutant mice. J Assoc Res Otolaryngol 1:161–171
LeBon L, Lee TV, Sprinzak D, Jafar-Nejad H, Elowitz MB (2014) Fringe proteins modulate Notch-ligand cis and trans interactions to specify signaling states. elife 3:e02950
Lewis, J. (1991). Rules for the production of sensory cells. Ciba Foundation Symposium 160 - Regeneration of Vertebrate Sensory Receptor Cells 25–53
Lewis RM, Hume CR, Stone JS (2012) Atoh1 expression and function during auditory hair cell regeneration in post-hatch chickens. Hear Res 289:74–85
Li S, Mark S, Radde-Gallwitz K, Schlisner R, Chin MT, Chen P (2008) Hey2 functions in parallel with Hes1 and Hes5 for mammalian auditory sensory organ development. BMC Dev Biol 8:20
Lin V, Golub JS, Nguyen TB, Hume CR, Oesterle EC, Stone JS (2011) Inhibition of Notch activity promotes nonmitotic regeneration of hair cells in the adult mouse utricles. J Neurosci 31:15329–15339
Lindsell CE, Boulter J, diSibio G, Gossler A, Weinmaster G (1996) Expression patterns of Jagged, Delta1, Notch1, Notch2, and Notch3 genes identify ligand-receptor pairs that may function in neural development. Mol Cell Neurosci 8:14–27

Liu M, Pereira FA, Price SD, Chu M, Shope C, Himes D, Eatock RA, Brownell WE, Lysakowski A, Tsai M-J (2000) Essential role of BETA2/NeuroD1 in development of the vestibular and auditory systems. Genes Dev 14:2839–2854

Liu Z, Owen T, Fang J, Zuo J (2012a) Overactivation of Notch1 signaling induces ectopic hair cells in the mouse inner ear in an age-dependent manner. PLoS One 7:e34123

Liu Z, Dearman JA, Cox BC, Walters BJ, Zhang L, Ayrault O, Zindy F, Gan L, Roussel MF, Zuo J (2012b) Age-dependent in vivo conversion of mouse Cochlear Pillar and Deiters' cells to immature hair cells by Atoh1 ectopic expression. J Neurosci 32:6600–6610

Liu Z, Brunskill E, Varnum-Finney B, Zhang C, Zhang A, Jay PY, Bernstein I, Morimoto M, Kopan R (2015) The intracellular domains of Notch1 and Notch2 are functionally equivalent during development and carcinogenesis. Development 142:2452–2463

Ma W-R, Zhang J (2015) Jag1b is essential for patterning inner ear sensory cristae by regulating anterior morphogenetic tissue separation and preventing posterior cell death. Development 142:763–773

Ma Q, Anderson DJ, Fritzsch B (2000) Neurogenin 1 null mutant ears develop fewer, morphologically normal hair cells in smaller sensory epithelia devoid of innervation. J Assoc Res Otolaryngol 1:129–143

Ma EY, Rubel EW, Raible DW (2008) Notch signaling regulates the extent of hair cell regeneration in the zebrafish lateral line. J Neurosci 28:2261–2273

Maass JC, Gu R, Basch ML, Waldhaus J, Lopez EM, Xia A, Oghalai JS, Heller S, Groves AK (2015) Changes in the regulation of the Notch signaling pathway are temporally correlated with regenerative failure in the mouse cochlea. Front Cell Neurosci 9

Maass JC, Gu R, Cai T, Wan Y-W, Cantellano SC, Asprer JST, Zhang H, Jen H-I, Edlund RK, Liu Z et al (2016) Transcriptomic analysis of mouse Cochlear supporting cell maturation reveals large-scale changes in Notch responsiveness prior to the onset of hearing. PLoS One 11:e0167286

Manderfield LJ, High FA, Engelka KA, Liu F, Li L, Rentschler S, Epstein JA (2012) Notch activation of Jagged1 contributes to the assembly of the arterial wall. Circulation 125:314–323

Mann ZF, Gálvez H, Pedreno D, Chen Z, Chrysostomou E, Żak M, Kang M, Canden E, Daudet N (2017) Shaping of inner ear sensory organs through antagonistic interactions between Notch signalling and Lmx1a. elife 6

Matei V, Pauley S, Kaing S, Rowitch D, Beisel KW, Morris K, Feng F, Jones K, Lee J, Fritzsch B (2005) Smaller inner ear sensory epithelia in Neurog1 null mice are related to earlier hair cell cycle exit. Dev Dyn 234:633–650

Millimaki BB, Sweet EM, Dhason MS, Riley BB (2007) Zebrafish atoh1 genes: classic proneural activity in the inner ear and regulation by Fgf and Notch. Development 134:295–305

Mizutari K, Fujioka M, Hosoya M, Bramhall N, Okano HJ, Okano H, Edge ASB (2013) Notch inhibition induces Cochlear hair cell regeneration and recovery of hearing after acoustic trauma. Neuron 77:58–69

Morrison A, Hodgetts C, Gossler A, Hrabé de Angelis M, Lewis J (1999) Expression of Delta1 and Serrate1 (Jagged1) in the mouse inner ear. Mech Dev 84:169–172

Morsli H, Choo D, Ryan A, Johnson R, Wu DK (1998) Development of the mouse inner ear and origin of its sensory organs. J Neurosci 18:3327–3335

Munnamalai V, Hayashi T, Bermingham-McDonogh O (2012) Notch prosensory effects in the mammalian cochlea are partially mediated by Fgf20. J Neurosci 32:12876–12884

Murata J, Tokunaga A, Okano H, Kubo T (2006) Mapping of notch activation during cochlear development in mice: implications for determination of prosensory domain and cell fate diversification. J Comp Neurol 497:502–518

Myat A, Henrique D, Ish-Horowicz D, Lewis J (1996) A chick homologue of< i> Serrate and its relationship with Notch and Delta homologues during central neurogenesis. Dev Biol 174:233–247

Nandagopal N, Santat LA, LeBon L, Sprinzak D, Bronner ME, Elowitz MB (2018) Dynamic ligand discrimination in the Notch signaling pathway. Cell 172:869–880.e19

Neves J, Kamaid A, Alsina B, Giraldez F (2007) Differential expression of Sox2 and Sox3 in neuronal and sensory progenitors of the developing inner ear of the chick. J Comp Neurol 503:487–500

Neves J, Parada C, Chamizo M, Giraldez F (2011) Jagged 1 regulates the restriction of Sox2 expression in the developing chicken inner ear: a mechanism for sensory organ specification. Development 138:735–744

Nichols DH, Pauley S, Jahan I, Beisel KW, Millen KJ, Fritzsch B (2008) Lmx1a is required for segregation of sensory epithelia and normal ear histogenesis and morphogenesis. Cell Tissue Res 334:339–358

Oesterle EC, Campbell S, Taylor RR, Forge A, Hume CR (2008) Sox2 and Jagged1 expression in Normal and drug-damaged adult mouse inner ear. J Assoc Res Otolaryngol 9:65–89

Ozeki M, Hamajima Y, Feng L, Ondrey FG, Schlentz E, Lin J (2007) Id1 induces the proliferation of cochlear sensory epithelial cells via the nuclear factor-κB/cyclin D1 pathway in vitro. J Neurosci Res 85:515–524

Pan W, Jin Y, Stanger B, Kiernan AE (2010) Notch signaling is required for the generation of hair cells and supporting cells in the mammalian inner ear. Proc Natl Acad Sci 107:15798–15803

Pan W, Jin Y, Chen J, Rottier RJ, Steel KP, Kiernan AE (2013) Ectopic expression of activated Notch or SOX2 reveals similar and unique roles in the development of the sensory cell progenitors in the mammalian inner ear. J Neurosci 33:16146–16157

Petrovic J, Formosa-Jordan P, Luna-Escalante JC, Abello G, Ibanes M, Neves J, Giraldez F (2014) Ligand-dependent Notch signaling strength orchestrates lateral induction and lateral inhibition in the developing inner ear. Development 141:2313–2324

Petrovic J, Gálvez H, Neves J, Abelló G, Giraldez F (2015) Differential regulation of Hes/Hey genes during inner ear development. Devel Neurobio 75:703–720

Pevny LH, Nicolis SK (2010) Sox2 roles in neural stem cells. Int J Biochem Cell Biol 42:421–424

Podgorski GJ, Bansal M, Flann NS (2007) Regular mosaic pattern development: a study of the interplay between lateral inhibition, apoptosis and differential adhesion. Theor Biol Med Model 4:43

Radosevic M, Robert-Moreno A, Coolen M, Bally-Cuif L, Alsina B (2011) Her9 represses neurogenic fate downstream of Tbx1 and retinoic acid signaling in the inner ear. Development 138:397–408

Raft S (2004) Suppression of neural fate and control of inner ear morphogenesis by Tbx1. Development 131:1801–1812

Raft S, Groves AK (2014) Segregating neural and mechanosensory fates in the developing ear: patterning, signaling, and transcriptional control. Cell Tissue Res 359:315–332

Raft S, Koundakjian EJ, Quinones H, Jayasena CS, Goodrich LV, Johnson JE, Segil N, Groves AK (2007) Cross-regulation of Ngn1 and Math1 coordinates the production of neurons and sensory hair cells during inner ear development. Development 134:4405–4415

Rauskolb C, Correia T, Irvine KD (1999) Fringe-dependent separation of dorsal and ventral cells in the Drosophila wing. Nature 401:476

Richardson RT, Atkinson PJ (2015) Atoh1 gene therapy in the cochlea for hair cell regeneration. Expert Opin Biol Ther 15:417–430

Riley BB, Chiang M, Farmer L, Heck R (1999) The deltaA gene of zebrafish mediates lateral inhibition of hair cells in the inner ear and is regulated by pax2. 1. Development 126:5669–5678

Rubel EW, Furrer SA, Stone JS (2013) A brief history of hair cell regeneration research and speculations on the future. Hear Res 297:42–51

Sánchez-Guardado LÓ, Puelles L, Hidalgo-Sánchez M (2013) Fgf10 expression patterns in the developing chick inner ear. J Comp Neurol 521:1136–1164

Sánchez-Guardado LÓ, Puelles L, Hidalgo-Sánchez M (2014) Fate map of the chicken otic placode. Development 141:2302–2312

Saravanamuthu SS, Gao CY, Zelenka PS (2009) Notch signaling is required for lateral induction of Jagged1 during FGF-induced lens fiber differentiation. Dev Biol 332:166–176

Schilder AGM, Su MP, Blackshaw H, Lustig L, Staecker H, Lenarz T, Safieddine S, Gomes-Santos CS, Holme R, Warnecke A (2019) Hearing protection, restoration, and regeneration: an overview of emerging therapeutics for inner ear and central hearing disorders. Otol Neurotol 40:559–570

Shaya O, Binshtok U, Hersch M, Rivkin D, Weinreb S, Amir-Zilberstein L, Khamaisi B, Oppenheim O, Desai RA, Goodyear RJ et al (2017) Cell-cell contact area affects notch signaling and notch-dependent patterning. Dev Cell 40:505–511.e6

Shida H, Mende M, Takano-Yamamoto T, Osumi N, Streit A, Wakamatsu Y (2015) Otic placode cell specification and proliferation are regulated by Notch signaling in avian development. Dev Dyn 244:839–851

Slowik AD, Bermingham-McDonogh O (2013) Hair cell generation by Notch inhibition in the adult mammalian cristae. J Assoc Res Otolaryngol 14:813–828

Steevens AR, Sookiasian DL, Glatzer JC, Kiernan AE (2017) SOX2 is required for inner ear neurogenesis. Sci Rep 7:4086

Steevens AR, Glatzer JC, Kellogg CC, Low WC, Santi PA, Kiernan AE (2019) SOX2 is required for inner ear growth and cochlear nonsensory formation before sensory development. Development 146:dev170522

Stone JS, Cotanche DA (2007) Hair cell regeneration in the avian auditory epithelium. Int J Dev Biol 51:633–647

Stone JS, Rubel EW (1999) Delta1 expression during avian hair cell regeneration. Development 126:961–973

Streit A (2002) Extensive cell movements accompany formation of the otic placode. Dev Biol 249:237–254

Tang LS, Alger HM, Pereira FA (2006) COUP-TFI controls Notch regulation of hair cell and support cell differentiation. Development 133:3683–3693

Tateya T, Imayoshi I, Tateya I, Ito J, Kageyama R (2011) Cooperative functions of Hes/Hey genes in auditory hair cell and supporting cell development. Dev Biol 352:329–340

Tateya T, Imayoshi I, Tateya I, Hamaguchi K, Torii H, Ito J, Kageyama R (2013) Hedgehog signaling regulates prosensory cell properties during the basal-to-apical wave of hair cell differentiation in the mammalian cochlea. Development 140:3848–3857

Taylor RR, Filia A, Paredes U, Asai Y, Holt JR, Lovett M, Forge A (2018) Regenerating hair cells in vestibular sensory epithelia from humans. elife 7:e34817

Togashi H, Kominami K, Waseda M, Komura H, Miyoshi J, Takeichi M, Takai Y (2011) Nectins establish a checkerboard-like cellular pattern in the auditory epithelium. Science 333:1144–1147

Troost T, Schneider M, Klein T (2015) A re-examination of the selection of the sensory organ precursor of the bristle sensilla of Drosophila melanogaster. PLoS Genet 11:e1004911

Tsai H, Hardisty RE, Rhodes C, Kiernan AE, Roby P, Tymowska-Lalanne Z, Mburu P, Rastan S, Hunter AJ, Brown SD et al (2001) The mouse slalom mutant demonstrates a role for Jagged1 in neuroepithelial patterning in the organ of Corti. Hum Mol Genet 10:507–512

Wang G-P, Chatterjee I, Batts SA, Wong HT, Gong T-W, Gong S-S, Raphael Y (2010) Notch signaling and Atoh1 expression during hair cell regeneration in the mouse utricle. Hear Res 267:61–70

Warchol ME, Lambert PR, Goldstein BJ, Forge A, Corwin JT (1993) Regenerative proliferation in inner ear sensory epithelia from adult guinea pigs and humans. Science 259:1619–1622

Warchol ME, Stone J, Barton M, Ku J, Veile R, Daudet N, Lovett M (2017) ADAM10 and γ-secretase regulate sensory regeneration in the avian vestibular organs. Dev Biol 428:39–51

Wilkerson BA, Chitsazan AD, VandenBosch LS, Wilken MS, Reh TA, Bermingham-McDonogh O (2019) Open chromatin dynamics in prosensory cells of the embryonic mouse cochlea. Sci Rep 9:1–15

Woods C, Montcouquiol M, Kelley MW (2004) Math1 regulates development of the sensory epithelium in the mammalian cochlea. Nat Neurosci 7:nn1349

Yamamoto N, Tanigaki K, Tsuji M, Yabe D, Ito J, Honjo T (2006) Inhibition of Notch/RBP-J signaling induces hair cell formation in neonate mouse cochleas. J Mol Med 84:37–45

Yamamoto N, Okano T, Ma X, Adelstein RS, Kelley MW (2009) Myosin II regulates extension, growth and patterning in the mammalian cochlear duct. Development 136:1977–1986

Yamamoto S, Charng W-L, Bellen HJ (2010) Chapter five - Endocytosis and intracellular trafficking of notch and its ligands. In: Kopan R (ed) Current topics in developmental biology. Academic Press, pp 165–200

Yamamoto N, Chang W, Kelley MW (2011) Rbpj regulates development of prosensory cells in the mammalian inner ear. Dev Biol 353:367–379

Yang H, Xie X, Deng M, Chen X, Gan L (2010) Generation and characterization of Atoh1-Cre knock-in mouse line. Genesis 48:407–413

Żak M, Klis SFL, Grolman W (2015) The Wnt and Notch signalling pathways in the developing cochlea: formation of hair cells and induction of regenerative potential. Int J Dev Neurosci 47:247–258

Zhang N, Martin GV, Kelley MW, Gridley T (2000) A mutation in the Lunatic fringe gene suppresses the effects of a Jagged2 mutation on inner hair cell development in the cochlea. Curr Biol 10:659–662

Zheng JL, Gao W-Q (2000) Overexpression of Math1 induces robust production of extra hair cells in postnatal rat inner ears. Nat Neurosci 3:580–586

Zheng JL, Shou J, Guillemot F, Kageyama R, Gao W-Q (2000) Hes1 is a negative regulator of inner ear hair cell differentiation. Development 127:4551–4560

Zine A, Aubert A, Qiu J, Therianos S, Guillemot F, Kageyama R, de Ribaupierre F (2001) Hes1 and Hes5 activities are required for the normal development of the hair cells in the mammalian inner ear. J Neurosci 21:4712–4720

Chapter 9
Notch Pathway and Inherited Diseases: Challenge and Promise

Jörg Reichrath and Sandra Reichrath

Abstract The evolutionary highly conserved Notch pathway governs many cellular core processes including cell fate decisions. Although it is characterized by a simple molecular design, Notch signaling, which first developed in metazoans, represents one of the most important pathways that govern embryonic development. Consequently, a broad variety of independent inherited diseases linked to defective Notch signaling has now been identified, including Alagille, Adams-Oliver, and Hajdu-Cheney syndromes, CADASIL (cerebral autosomal-dominant arteriopathy with subcortical infarcts and leukoencephalopathy), early-onset arteriopathy with cavitating leukodystrophy, lateral meningocele syndrome, and infantile myofibromatosis. In this review, we give a brief overview on molecular pathology and clinical findings in congenital diseases linked to the Notch pathway. Moreover, we discuss future developments in basic science and clinical practice that may emerge from recent progress in our understanding of the role of Notch in health and disease.

Keywords Notch · Notch signaling · Notch pathway · Embryonic development · Jagged · Delta-like ligand

Abbreviations

AD	Autosomal dominant
ALGS	Alagille syndrome
ARHGAP31	RhoGTPase-activating protein 31
BAV	Bicuspid aortic valve
BMP	Bone morphogenetic protein
CADASIL	Cerebral autosomal-dominant arteriopathy with subcortical infarcts and leukoencephalopathy

J. Reichrath (✉) · S. Reichrath
Department of Dermatology, The Saarland University Hospital, Homburg, Germany
e-mail: joerg.reichrath@uks.eu

J. Reichrath, S. Reichrath (eds.), *Notch Signaling in Embryology and Cancer*, Advances in Experimental Medicine and Biology 1218,
https://doi.org/10.1007/978-3-030-34436-8_9

CAVD	Calcific aortic valve disease
CHD	Congenital heart disease
cKO	Conditional knockout
CNS	Central nervous system
Dll	Delta-like canonical Notch ligand
DOCK	Dedicator of cytokinesis
E	Embryonic day
EGF	Epidermal growth factor
EMT	Epithelial-to-mesenchymal transition
ENU	*N*-Ethyl *N*-nitrosourea
EOGT	EGF domain-specific O-linked N-acetylglucosamine transferase
FGF	Fibroblast growth factor
HCS	Hajdu-Cheney syndrome
Hes	Hairy and enhancer of split
HLHS	Hypoplastic left heart syndrome
IM	Infantile myofibromatosis
Jag	Jagged
KO	Knockout
LMS	Lateral meningocele syndrome
LOF	Loss of function
LW	Lateral wall
MET	Mesenchymal-to-epithelial transition
NEPs	Neuroepithelial cells
NICD	Notch intracellular domain
NRR	Negative regulatory region
NSCs	Neural stem cells
OMIM	Online Mendelian Inheritance in Man
PEST sequence	Peptide sequence that is rich in proline (P), glutamic acid (E), serine (S), and threonine (T)
RBPJ	Recombination signal binding protein for immunoglobulin kappa J region
TAA	Thoracic aortic aneurysms
TOF	Tetralogy of Fallot
VSD	Ventricular septal defect
vSMC	Vascular smooth muscle cell
Wnt	Wingless

Introduction

The Notch pathway, which regulates many cellular core processes including cell fate decisions, is characterized by functional diversity, although its design is quite simple (Andersson et al. 2011). It first developed during evolution in metazoans (Gazave et al. 2009; Richards and Degnan 2009) and was first discovered in

Drosophila melanogaster. In the last decades, a huge mountain of new scientific information has clearly demonstrated that Notch signaling represents one of the most important pathways that govern embryogenesis in animals and humans. As outlined elsewhere in this book (Reichrath and Reichrath 2020a), the tale that earned the gene the name *Notch* began over a century ago, when the American Scientist John S. Dexter discovered at Olivet College (Olivet, Michigan, USA) the characteristic notched-wing phenotype (a nick or notch in the wingtip) in his stock of mutant fruit flies *Drosophila melanogaster* (Dexter 1914). The alleles responsible for this phenotype were identified 3 years later at Columbia University (New York City, New York, USA) by another American scientist, Thomas Hunt Morgan (Morgan 1917), who discovered various mutant loci in the chromosomes of these fruit flies that were associated with several distinct notched-wing phenotypes. Although the majority of them was lethal, these alleles were associated with the characteristic phenotype with a nick in the wingtip and bristle phenotype specifically in female fruit flies, suggesting an association of these alleles with the X chromosome (Morgan 1928). Notably, this discovery and similar investigations that supported the chromosomal theory of inheritance earned Thomas Hunt Morgan in 1933 the Nobel Prize in physiology/medicine. In subsequent decades, despite the extensive research on the *Notch* locus, researchers struggled to identify the function for the *Notch* gene due to the lethality early in embryogenesis and broad phenotypic consequences of Notch mutants.

In the following years, many additional alleles were identified, which were associated with the Notch phenotype (Morgan 1928). This observation was finally confirmed in the laboratories of Spyros Artavanis-Tsakonas and Michael W. Young more than half a century later by cloning and sequencing of the mutant *Notch* locus (Wharton et al. 1985; Kidd et al. 1986).

Although this pathway is characterized by a relatively simple design, Notch signaling exerts highly versatile functions (Andersson et al. 2011), playing multiple roles, both in development and in adult tissue homeostasis, including keeping precursor and stem cells in a non-differentiated state, having the ability to activate cell proliferation and to regulate various other important cell fate decisions (reviewed by Kopan and Ilagan 2009; Reichrath and Reichrath 2020). Until today, a huge mountain of studies – ranging from the molecular and functional elucidation of the Notch pathway (reviewed in Bray 2016; Kopan and Ilagan 2009) to the generation of knockouts in model organisms and the discovery of Notch genes mutated in humans (Gridley 2003) – has confirmed an essential role for Notch signaling in human development. Consequently, a broad variety of independent inherited disorders linked to defective Notch signaling has now been identified. A major breakthrough in the investigation of these congenital diseases was the linkage analysis-based discovery of heterozygous *NOTCH3* mutations on chromosome 19 in patients diagnosed with CADASIL (cerebral autosomal-dominant arteriopathy with subcortical infarcts and leukoencephalopathy, an autosomal-dominant hereditary stroke disorder resulting in vascular dementia) that was reported by Joutel et al. in 1996. In the following year, two groups independently reported that mutations in *JAG1*, located within chromosome 20p12, cause Alagille syndrome (Li et al. 1997; Oda et al. 1997).

Since then, several other inherited disorders, involving pathological embryogenesis in multiple tissues, including Adams-Oliver and Hajdu-Cheney syndromes, have now convincingly been linked to defective Notch signaling. The fact that many of these congenital diseases are rare (with prevalences of just a few per 100,000) not only represents serious hurdles to studying the impact of these genes in humans (Mašek and Andersson 2017). It has been concluded that it also underlines the crucial importance of Notch signaling for human survival (Mašek and Andersson 2017). Notably, the generation of knockout mice and investigations using other animal models have in recent years resulted in the generation of a huge mountain of new scientific findings regarding the function of specific Notch components in human development and disease (Mašek and Andersson 2017). An important example of these animal models are *Jag1* mouse mutants generated in *N*-ethyl *N*-nitrosourea (ENU) mutagenesis screens (ENU is an alkylating agent, acting as a potent mutagen, with preference for A→T base transversions and also for AT→GC transitions, but also causing GC→AT transitions), such as Ozzy (W167R, Delta/Serrate LAG-2 domain, Vrijens et al. 2006), Headturner (G289D, EGF-like repeats, Kiernan et al. 2001), Slalom (P269S, EGF-like repeats, Tsai et al. 2001), and Nodder (H268O, EGF-like repeats; Hansson et al. 2010), that harbor mutations that cluster in the N-terminal missense-mutation hotspot typically found in Alagille syndrome. Interestingly, recent "big data" analyses of whole-exome and whole-genome sequencing have now revealed the presence or absence of mutations in Notch components in the human population, confirming that specific Notch components are essential to species fitness and presenting exciting future avenues of research (Mašek and Andersson 2017). In this review, we give a short overview on molecular pathology and clinical findings in congenital diseases linked to the Notch pathway. Moreover, we briefly discuss the emerging role of Notch as a promising therapeutic target.

Defective Notch Signaling and Embryogenesis: Congenital Heart Disease (CHD) as an Example for the Broad Variety of Pathophysiological Consequences in Individual Tissues

Bicuspid aortic valve (BAV) represents the most common congenital heart defect, affecting 1–2% of the general population (reviewed in Meester et al. 2019). While in many cases, BAV exerts no or only mild clinical symptoms and remains undetected, it can also result in severe cardiovascular complications including calcific aortic valve disease (CAVD) (Garg et al. 2005; reviewed in Meester et al. 2019), coarctation, stenosis, and valve dysfunction (Michelena et al. 2014; reviewed in Meester et al. 2019). Moreover, it has been reported that in at least 20% of affected patients, BAV is associated with the development of thoracic aortic aneurysms (TAA) (Gillis et al. 2017; reviewed in Meester et al. 2019) and that BAV-associated complications are associated with significantly increased mortality rates (reviewed in Meester et al. 2019). Despite the fact that BAV is common, very few genes have been linked to this condition and the hitherto known genes only explain disease in a minority of patients (reviewed in Meester et al. 2019). Garg et al. published in 2005 the first study reporting the association of *NOTCH1* mutations in congenital heart disease (CHD) (Garg et al. 2005). In that investigation, truncating *NOTCH1* mutations were detected in members of two separate families that suffered from various aortic and cardiac defects, including BAV, CAVD, aortic stenosis, aortic insufficiency, TAA, tetralogy of Fallot (TOF), ventricular septal defect (VSD), mitral atresia, hypoplastic left ventricle, and double-outlet right ventricle (Garg et al. 2005; reviewed in Meester et al. 2019). In the following years, many other investigations reported associations between variations in *NOTCH1* and BAV, BAV/TAA, hypoplastic left heart syndrome (HLHS), aortic valve stenosis, and coarctation (reviewed in Meester et al. 2019). However, it has to be noted that most of the mutations detected in these investigations are missense mutations, which do not replace or create critical cysteine or other conserved residues in the EGF-like domains, of which the vast majority has also been reported in public databases (e.g., gnomAD, reviewed in Meester et al. 2019), sometimes with high frequencies. Therefore, it has been concluded that the causal relationship of these missense mutations with CHD is not as convincing as the causal relationship of the loss-of-function (LOF) mutations from the initial report (Garg et al. 2005; reviewed in Meester et al. 2019).

Interestingly, a large-scale screening investigation of 428 patients with left-sided CHD (LS-CHD), confined to aortic valve stenosis, BAV, coarctation of the aorta, and HLHS reported in 2016 the presence of 14 different *NOTCH1* mutations, including splicing mutations, truncating mutations, and a whole gene deletion (Kerstjens-Frederikse et al. 2016; reviewed in Meester et al. 2019). Interestingly, a specific frameshift mutation detected in this investigation (p.Ser2486Leufs∗21) is located within the last exon and is therefore predicted to escape nonsense-mediated decay (NMD) of the mutant mRNA transcript (Kerstjens-Frederikse et al. 2016; reviewed in Meester et al. 2019). Consequently, it has been hypothesized that this mutation should result in a dominant negative effect. Notably, this effect should be different from the other truncating mutations reported in this investigation, for which it was speculated that haploinsufficiency (HI) would be the most likely mechanism of disease (reviewed in Meester et al. 2019). Moreover, it has been reported that 18% of mutation carriers suffered from right-sided CHD (RS-CHD) or conotruncal heart disease, revealing that the observed CHD has both a left- and right-sided localization (Kerstjens-Frederikse et al. 2016; reviewed in Meester et al. 2019). Furthermore, in 10% of *NOTCH1* mutation carriers, TAA has been detected (reviewed in Meester et al. 2019). Interestingly, familial segregation revealed that 25% of mutation carriers were asymptomatic, indicating a significantly decreased penetrance (Kerstjens-Frederikse et al. 2016; reviewed in Meester et al. 2019).

However, a recent investigation could not confirm these findings, reporting conflicting data (Gillis et al. 2017; reviewed in Meester et al. 2019). Results of that study, which investigated 441 patients with BAV/TAA, indicated a possible protective role for *NOTCH1* variants, as missense/splicing variants were observed more frequently amongst control populations compared to the BAV/TAA cohort (reviewed in Meester et al. 2019). However, it has to be noted that, as the authors state, sample selection bias might have contributed to this observation, as *NOTCH1* variants appear to associate with early and severe valve calcification and seem to be enriched in families with highly penetrant BAV but far lower penetrance of TAA (Gillis et al. 2017; reviewed in Meester et al. 2019).

To study the role of this receptor, several Notch1 mouse models have been generated. Homozygous knockout of Notch1 leads to embryonic lethality due to vascular defects, indicating an essential role for Notch signaling in early cardiovascular development (Krebs et al. 2000; reviewed in Meester et al. 2019). Heterozygosity of Notch1 on a Nos3-null background, a model previously known for the development of BAV (Lee et al. 2000), is characterized by high penetrance of BAV (Bosse et al. 2013). It has been shown that endothelial-specific loss of Notch1 contributes to the development of BAV (Koenig et al. 2016; Wang et al. 2017; reviewed in Meester et al. 2019). Notably, endothelial Dll4 is required for epithelial-to-mesenchymal transition (EMT), a highly coordinated process characterized by the detachment of endocardial cells in the atrioventricular canal and outflow tract and their transition to mesenchyme cells of the endocardial cushions (reviewed in Meester et al. 2019). On the other hand, endocardial Jag1 has been demonstrated to be essential for proper cushion formation at post-EMT stages (MacGrogan et al. 2016; reviewed in Meester et al. 2019). Interestingly, calcification studies of the aortic valves revealed that immortalized Notch1$^{+/-}$ aortic valve interstitial cells resemble a myofibroblast-like phenotype, expressing higher amounts of mediators of dystrophic calcification (Chen et al. 2015; reviewed in Meester et al. 2019). Moreover, recent investigations have convincingly demonstrated that a heterozygous loss of Notch1 (Notch1$^{+/-}$) causes the development of TAA on a 129SV background, a phenomenon not observed on a mixed background (C59Bl6, 129SV, BTBR) (Koenig et al. 2017; reviewed in Meester et al. 2019).

In summary, there is at present convincing evidence that truncating mutations in *NOTCH1* cause a wide range of CHD, characterized by incomplete penetrance and variable expression. In contrast, the causative potential of NOTCH1 missense variants for the pathogenesis of CHD is less convincing and needs to be clarified in future investigations (reviewed in Meester et al. 2019).

Alagille Syndrome (ALGS)

Molecular Biology and Genetics of Alagille Syndrome (ALGS)

This multisystemic inherited disease was first described in 1975 by Alagille et al. (1975). It's prevalence was estimated to be 1 in 70,000. However, this is likely an underestimation, as it was based on the presence of neonatal liver disease, and it was later discovered that a highly variable phenotype is present (Table 9.1, reviewed in Meester et al. 2019). Alagille syndrome (ALGS) is caused by loss-of-function mutations in either JAG1 (OMIM 601920, cytogenetic location: 20p12.2; 94–96%

Table 9.1 Molecular pathology and clinical findings in genetic disorders linked to the Notch signaling pathway

Genetic disorder	Molecular pathology	Clinical findings	References
Alagille syndrome (ALGS)	Loss-of-function mutations in *JAG*1 (cytogenetic location: 20p12.2, OMIM 601920) or *NOTCH*2 (cytogenetic location: 1p12, OMIM 600275)	Five hallmarks: (1) Characteristic facial features (include a prominent forehead, pointed chin, and deep-set eyes) (2) Posterior embryotoxon (a distinct eye defect) (3) A broad variety of heart defects (that may range from pulmonary stenosis to tetralogy of Fallot) (4) Vertebral defects (including typical butterfly vertebrae) (5) Jaundice/cholestasis (resulting from intrahepatic bile duct paucity)	Alagille et al. (1975), Emerick et al. (1999), Kamath et al. (2004, 2013), Mašek and Andersson (2017), Zanotti and Canalis (2016)
		In addition a significant number of patients show growth retardation (50–90%), renal symptoms (40%), and/or vascular structural anomalies and bleeds (10–25%) The severity of the clinical phenotype may vary greatly between members of the same family, with a broad variety of symptoms ranging from unnoticed and/or mild as to severe heart and/or liver diseases that require transplantation	
Adams-Oliver syndrome	Mutations in: *DLL4* (OMIM 605185, cytogenetic location: 15q15.1; autosomal dominant), *NOTCH*1 (OMIM 190198, cytogenetic location: 9q34.3; autosomal dominant), *RBPJ* (OMIM 147183, cytogenetic location: 4q15.2),	Three hallmarks: (1) Terminal transverse limb malformations (can resemble amputations, patients may also have syndactyly) (2) Partial absence of skin (named aplasia cutis congenita, predominantly found in the skull region) (3) Partial absence of skull bones	Hassed et al. (2012), Isrie et al. (2014), Lehman et al. (2014), Mašek and Andersson (2017), Meester et al. (2015), Patel et al. (2004), Shaheen et al. (2011),
	EOGT (OMIM 614789, cytogenetic location: 3p14.1; autosomal recessive), *ARHGAP*31 (OMIM 610911, cytogenetic location: 3q13.2-3q13.33; autosomal dominant), and *DOCK*6 (OMIM 614194, cytogenetic location: 19p13.2; autosomal recessive)	It has been speculated that most symptoms of Adams-Oliver syndrome are caused by impaired circulation. Additional symptoms may include congenital heart defects (around 23%), vascular anomalies (including dilated surface blood vessels, which result in a marbled appearance of affected skin areas, termed cutis marmorata telangiectatica), pulmonary or portal hypertension, and retinal hypervascularization The severity of the clinical phenotype may vary greatly between members of the same family	Southgate et al. (2011, 2015), Stittrich et al. (2014), Sukalo et al. (2015a, b), Swartz et al. (1999), Zanotti and Canalis (2016)

(continued)

Table 9.1 (continued)

Genetic disorder	Molecular pathology	Clinical findings	References
Hajdu-Cheney syndrome (HCS)	Nonsense mutations (point mutations) in exon 34 of *NOTCH2* (OMIM 102500, cytogenetic location: 1p12; leading to the creation of a stop codon and the premature termination of the protein product upstream the PEST domain (truncated protein)). The NOTCH2 protein that is synthesized is stable and active. Since the PEST domain contains sequences necessary for the ubiquitinylation and degradation of Notch in the proteasome, the mutations lead to a stable protein and persistence of NOTCH2 signaling since all sequences required for the formation of the Notch transcriptional complex are upstream the PEST domain and are therefore preserved	Typically facial dysmorphism, synophrys, epicanthal folds, and relatively short distance between the eyes. Additionally, thick eyebrows extend toward the midline; malar hypoplasia, a long and smooth philtrum, micrognathia, a flattened nasal bridge that then becomes broad, and other facial features (including coarse and a short neck) may be present Craniofacial developmental defects may include wormian bones, open sutures, platybasia, and basilar invagination Additional clinical signs may include cardiovascular defects (including patent ductus arteriosus, septal defects, and mitral and aortic valve abnormalities associated with valvular insufficiency or stenosis), polycystic kidneys (about 10% of HCS patients – serpentine fibula-polycystic kidney syndrome may represent the same disorder as HCS), abnormal dental eruptions, tooth decay with premature loss of teeth, short stature, generalized and local joint hypermobility, and vertebral abnormalities (including fractures, kyphosis and scoliosis, and long-bone deformities)	Hajdu and Kauntze (1948), Cheney (1965), Mašek and Andersson (2017), Zanotti and Canalis (2016)
		A characteristic feature is acroosteolysis of the distal phalanges of fingers and toes, associated with a local inflammatory reaction, pain, and swelling, a process that may cause the loss of the distal phalanges, thereby shortening hands and feet. Although the distal phalangeal osteolytic lesions suggest increased localized bone resorption, there is no information on the mechanisms responsible for the generalized osteoporosis. Iliac crest biopsies show decreased trabecular bone, normal or increased bone remodeling, and normal or decreased bone formation. It is not known whether the osteoblast/osteocyte or the osteoclast is the cell responsible for the presumed change in bone turnover. In osteoclast precursors, Notch2, but not Notch1, induces nuclear factor of T-cell 1 transcription and osteoclastogenesis. Whether this mechanism operates in HCS is not known. Notably, in HCS, the mechanisms that underlie the acroosteolysis are not considered to be the same as those that cause the generalized bone loss and fractures that may be present	

		High clinical variability. Typically abnormalities of craniofacial development at a young age in childhood that evolve as the person matures. These craniofacial developmental defects represent serious manifestations because they can cause severe neurological complications, which may even result in central respiratory arrest and sudden death	
		Bisphosphonate therapy (alendronate and pamidronate) alone or in combination with anabolic therapy with teriparatide has been attempted for treatment of skeletal manifestations, but there is no clear evidence that either therapy is beneficial	
Cerebral autosomal-dominant arteriopathy with subcortical infarcts and leukoencephalopathy (CADASIL)	Heterozygous mutations in *NOTCH3* (OMIM 125310, autosomal dominant), clustering in four "hotspots," including exon 4 and exon 11. Most mutations do not cause loss of function, but exert neomorphic or toxic effects (resulting in deletion or addition of a cysteine residue in the characteristic EGF-like repeats located in the ECD of this transmembrane receptor protein, thereby causing aggregation of the NOTCH3 ECD into extracellular deposits of granular osmiophilic material (GOM))	Notch signaling is essential for vascular physiology. Clinically, patients typically suffer from multiple ischemic strokes caused by an arteriopathy that shows breakdown of vascular smooth muscle cells (vSMCs) – a cell population in which NOTCH3 is highly expressed – and vascular dementia. Diagnostic hallmark: cerebral white matter lesions detected by magnetic resonance imaging (MRI)	Abou Al-Shaar et al. (2016), Domenga et al. (2004), Wang et al. (2012), Joutel et al. (1996, 1997, 2000), Liem et al. (2008), Mašek and Andersson (2017), Pippucci et al. (2015), Ragno et al. (2013), Soong et al. (2013), Tuominen et al. (2001), Vinciguerra et al. (2014)
Early-onset arteriopathy with cavitating leukodystrophy	Homozygous c.C2898A (p.C966*) null mutation in *NOTCH3* abolishing *NOTCH3* expression, causing NOTCH3 signaling impairment, and downregulating. NOTCH3 targets acting in the regulation of arterial tone (KCNA5) or expressed in the vasculature (CDH6)	A patient with childhood-onset arteriopathy, cavitating leukoencephalopathy with cerebral white matter abnormalities presented as diffuse cavitations, multiple lacunar infarctions, and disseminated microbleeds. Vessels were characterized by smooth muscle degeneration as in CADASIL, but without deposition of granular osmiophilic material (GOM), the CADASIL hallmark. The heterozygous parents displayed similar but less dramatic trends in decrease in the expression of *NOTCH3* and its targets, as well as in vessel degeneration	Mašek and Andersson (2017), Pippucci et al. (2015)

(continued)

Table 9.1 (continued)

Genetic disorder	Molecular pathology	Clinical findings	References
Lateral meningocele syndrome (LMS, also termed Lehman syndrome)	A very rare inherited disease associated with heterozygous truncating *NOTCH3* mutations (OMIM 130720), which cluster into the last coding exon, resulting in premature termination of the protein and truncation of the negative regulatory proline-glutamate-serine-threonine-rich PEST domain	Skeletal disorder that has phenotypic overlap with Hajdu-Cheney syndrome, associated with facial anomalies, hypotonia, and/or meningocele-related neurologic malfunction	Gripp et al. (2015), Mašek and Andersson (2017), Yu et al. (2019), Zanotti and Canalis (2016)
	Truncated *NOTCH3* may cause gain-of-function through decreased clearance of the active intracellular product, resembling *NOTCH2* mutations in clinically related Hajdu-Cheney syndrome and contrasting *NOTCH3* missense mutations causing CADASIL	The characteristic lateral meningoceles are the clinical hallmarks of the disease. Being in general most severe in the lower spine, they represent the severe end of the dural ectasia spectrum	
	In a mouse model of LMS ($Notch3^{tm1.1Ecan}$), cancellous bone osteopenia was no longer detected after intraperitoneal administration of antibodies directed against the negative regulatory region (NRR) of Notch3. In that study, anti-Notch3 NRR antibody suppressed expression of Hes1, Hey1, and Hey2 (Notch target genes) and decreased Tnfsf11 (receptor activator of NF kappa B ligand) messenger RNA in $Notch3^{tm1.1Ecan}$ osteoblast cultures. This study indicates that cancellous bone osteopenia of $Notch3^{tm1.1Ecan}$ mutants can be reversed by anti-Notch3 NRR antibodies, thereby opening new avenues for treatment of bone osteopenia in LMS patients	Additional clinical features may include facial abnormality (hypertelorism and telecanthus, high-arched eyebrows, ptosis, midfacial hypoplasia, micrognathia, high and narrow palate, low-set ears, and a hypotonic appearance), connective tissue abnormalities (hyperextensibility, hernias, and scoliosis), aortic dilation, a high-pitched nasal voice, wormian bones, and osteolysis	

Infantile myofibromatosis-2 (IM-2)	Heterozygous c.4556T>C (p. Leu1519Pro) mutation in *NOTCH3* (cytogenetic location 19p13, OMIM 615293). Until today, only one family has been reported	Typically, nonmetastasizing mesenchymal tumors develop in the skin, muscle, bone, and viscera (in most cases in neonates or infants under 24 months of age, with few reports of adult onset). Solitary, multicentric, or generalized forms of the disease have been reported. Most patients present with a single (solitary form) or multiple (generalized form) cutaneous nodules. The multicentric form is characterized by mesenchymal tumors in the skin, subcutaneous tissues, muscles, and bone. Prognosis is in general good (no metastases, in many cases regression of the tumor over a period of 12–18 months). In contrast, the generalized form has an unfavorable course, being associated with visceral involvement and having a 76% mortality rate from cardiopulmonary or gastrointestinal complications	Martignetti et al. (2013), Mašek and Andersson (2017), Zanotti and Canalis (2016)
	While most cases of IM appear to be sporadic, occurrence within families across multiple generations indicates an autosomal-dominant inheritance pattern, and autosomal-recessive modes of inheritance have also been suggested. Mutations in the *PDGFRB* and *NOTCH3* genes were recently identified in patients with IM. In eight of nine families, one of two disease-causing mutations, c.1978C>A (p.Pro660Thr) and c.1681C>T (p.Arg561Cys), were detected in *PDGFRB*. Interestingly, one family did not have either of these *PDGFRB* mutations on chromosome 5q32, but all affected individuals had a heterozygous c.4556T>C (p. Leu1519Pro) mutation in *NOTCH3*	There is no standard therapy and treatment options vary widely. Solitary and even multicentric cutaneous and subcutaneous lesions without visceral involvement often regress spontaneously. However, calcification and atrophic scars frequently remain after spontaneous regression. Extensive surgery and chemotherapy were reported to be beneficial for multicentric disease	

Abbreviations: *ALGS* Alagille syndrome, *ARHGAP31* RhoGTPase-activating protein 31, *CADASIL* cerebral autosomal-dominant arteriopathy with subcortical infarcts and leukoencephalopathy, *cKO* conditional knockout, *Dll* delta-like canonical Notch ligand, *DOCK* dedicator of cytokinesis, *ECD* extracellular domain, *EOGT* EGF domain-specific O-linked N-acetylglucosamine transferase, *GOF* gain of function, *HCS* Hajdu-Cheney syndrome, *Hes* hairy and enhancer of split, *IM* infantile myofibromatosis, *jag* jagged, *KO* knockout, *LMS* lateral meningocele syndrome, *NICD* Notch intracellular domain, *NSC* neural stem cell, *OMIM* Online Mendelian Inheritance in Man, *PEST sequence* peptide sequence that is rich in proline (P), glutamic acid (E), serine (S), and threonine (T), *RBPJ* recombination signal binding protein for immunoglobulin kappa J region, *Tnfsf11* receptor activator of NF kappa B ligand, *vSMC* ventricular smooth muscle cell of affected patients) or NOTCH2 (OMIM 600275, cytogenetic location: 1p12; 1–2% of affected patients) (Table 9.1, Descartes et al. 2014; Gray et al. 2012; Han et al. 2015; Isidor et al. 2011a, b; Kamath et al. 2012; Li et al. 1997; Majewski et al. 2011; McDaniell et al. 2006; Narumi et al. 2013; Oda et al. 1997; Simpson et al. 2011). Because ALGS is inherited in an autosomal-dominant pattern, one copy of the altered gene will cause the disorder (Mašek and Andersson 2017). *JAG1 m*utations can represent deletions, truncations, splice site, nonsense, or missense. It was previously suggested that mutations linked to ALGS could occur anywhere in the coding domains (CDS) for *JAG1*, but analysis of 87 missense mutations revealed that the deleterious mutations that are associated with ALGS predominantly cluster in the N-terminal region of JAG1 and with two other smaller sub-clusters: one in EGF-like repeats 11–12 and one in the von Willebrand factor type C/Jagged Serrate domain (also known as the cysteine-rich domain) (Mašek and Andersson 2017). Two characteristic types of Alagille *NOTCH2* mutations have been detected in ALGS, either abrogating cysteines in the ligand-binding EGF-like repeats or arginines in the ankyrin repeats (Descartes et al. 2014; Gray et al. 2012; Han et al. 2015; Isidor et al. 2011a, b; Majewski et al. 2011; Narumi et al. 2013; Simpson et al. 2011).

Clinical Hallmarks of Alagille Syndrome

ALGS has been defined as a multisystemic inherited disease characterized by the presence of bile duct paucity in combination with three out of five major criteria, including cholestatic liver disease, cardiac anomalies, ocular abnormalities, skeletal defects, and characteristic craniofacial features (Table 9.1, Alagille et al. 1987; reviewed in Meester et al. 2019). More in detail, the five clinical hallmarks of this inherited disease may include the following symptoms: (1) characteristic facial features that include a prominent forehead, pointed chin, and deep-set eyes, (2) posterior embryotoxon (a distinct eye defect), (3) a broad variety of heart defects that may range from pulmonary stenosis to tetralogy of Fallot, (4) vertebral defects (including typical butterfly vertebrae), and (5) jaundice/cholestasis resulting from intrahepatic bile duct paucity. In addition to these diagnostic hallmarks, a significant number of patients show growth retardation (50–90%) (Alagille et al. 1975, 1987; Emerick et al. 1999), have renal symptoms (40%) (Kamath et al. 2013), and/or present with vascular structural anomalies and bleeds (10–25%) (Emerick et al. 1999; Kamath et al. 2004).

The vast majority of ALGS patients present with liver disease, including mild cholestasis, jaundice, and pruritis, that can progress to liver failure within the first 3 months of life (reviewed in Meester et al. 2019). ALGS-associated liver damage can result in a yellowish tinge in the skin and the whites of the eyes (jaundice), itching (pruritus), pale stools (acholia), an enlarged liver (hepatomegaly), an enlarged spleen (splenomegaly), and deposits of cholesterol in the skin (xanthomas) (reviewed in Meester et al. 2019). Additionally, too few bile ducts may be present (bile duct paucity) or, in some cases, bile ducts may be completely lacking (biliary atresia) (reviewed in Meester et al. 2019). It was reported that bile duct paucity may lead to reduced absorption of fat and vitamins (A, D, E, and K), which may cause rickets or a failure to thrive in children (reviewed in Meester et al. 2019). Around 15% of affected patients will develop liver cirrhosis in the course of their disease, and hepatocellular cancer has been observed in some cases (Sijmons 2008).

Approximately 94% of ALGS patients suffer from cardiac manifestations, which also vary widely between affected individuals and range from benign heart murmurs to significant structural malformations (reviewed in Meester et al. 2019), such as tetralogy of Fallot. Other ALGS-associated heart defects include pulmonary stenosis, overriding aorta, ventricular septal defect, atrial septal defects, patent ductus arteriosus, coarctation of the aorta, and right ventricular hypertrophy (reviewed in Meester et al. 2019). Without therapy, tetralogy of Fallot mortality rates range from 70% to 95%, by age 10 to age 40, respectively (reviewed in Meester et al. 2019). In ALGS patients, complete surgical repair significantly increases both longevity and quality of life (reviewed in Meester et al. 2019).

Other presentations of ALGS include a characteristic butterfly shape of one or more of the bones of the spinal column, distinct eye defects (such as posterior embryotoxon and pigmentary retinopathy), and narrowed pulmonary arteries that may increase pressure on the right heart valves (reviewed in Meester et al. 2019). Many patients with ALGS have similar facial features, including a broad, prominent forehead, deep-set eyes, and a small pointed chin (reviewed in Meester et al. 2019). Additionally, the kidneys and the central nervous system can also be affected (reviewed in Meester et al. 2019).

Recent studies have indicated that ALGS is accompanied by reduced penetrance and markedly variable expression (reviewed in Meester et al. 2019). Importantly, familial segregation analyses revealed a substantial number of mutation carriers that did not fulfil all clinical diagnostic criteria of ALGS (reviewed in Meester et al. 2019). Consequently, it has been hypothesized that the clinical diagnostic criteria might be too stringent and more emphasis should be placed on the molecular identification of pathogenic variants in this disease (reviewed in Meester et al. 2019).

In ALGS, the severity of the clinical phenotype may vary greatly between members of the same family, with a broad variety of symptoms ranging from unnoticed and/or mild as to severe heart and/or liver diseases that require transplantation (Gunadi et al. 2019). Although it is in general difficult to predict the prognosis of an individual patient suffering from ALGS, several indicators of earlier death have been reported (Emerick et al. 1999). Following liver transplantation in ALGS, the effect of long-term immunosuppression on other affected systems has not been evaluated well until today. Therefore long-term posttransplant prospective studies are urgently needed to address these issues (Singh and Pati 2018, reviewed in Meester et al. 2019).

Adams-Oliver Syndrome: Highlighting the Impact of NOTCH1, DLL4, and RBPJk for Human Embryogenesis

Historical Considerations

Historically, it has to be noted that until 1945, when Clarence Paul Oliver and Forrest H. Adams reported their thoughtful observations (Adams and Oliver 1945), newborns presenting with underdeveloped upper or lower extremities were considered as having congenital amputations, defects that were attributed to amniotic band or umbilical cord constriction of the extremities (amniotic band syndrome) (reviewed in Mašek and Andersson 2017; Meester et al. 2019). In 1945, Clarence Paul Oliver and Forrest H. Adams described a patient with anomalies in the feet and one hand and also a denuded area of the scalp, with a thinner skull. Most importantly, they demonstrated that many family members had similar symptoms and speculated that the condition was hereditary (Adams and Oliver 1945). Since then, the diagnosis, genetics, and underlying biology of Adams-Oliver syndrome, as it has come to be known, has become more complex (reviewed in Mašek and Andersson 2017; Meester et al. 2019). The condition presenting with underdeveloped upper or lower extremities is now known as terminal transverse limb deficiencies (reviewed in Mašek and Andersson 2017; Meester et al. 2019).

Molecular Biology and Genetics of Adams-Oliver Syndrome

Adams-Oliver syndrome represents a rare genetic disorder that can be autosomal dominant or autosomal recessive or caused by de novo mutations (reviewed in Mašek and Andersson 2017; Meester et al. 2019). It has been linked to mutations in several different genes, including *DLL4* (OMIM 605185, cytogenetic location: 15q15.1) and *NOTCH*1 (OMIM 190198, cytogenetic location: 9q34.3), as well as in RBPJ (OMIM 147183, cytogenetic location: 4q15.2), *EOGT* (OMIM 614789, cytogenetic location: 3p14.1), *ARHGAP*31 (OMIM 610911, cytogenetic location: 3q13.2–3q13.33), and *DOCK*6 (OMIM 614194, cytogenetic location: 19p13.2) (reviewed in Mašek and Andersson 2017; Meester et al. 2019). While mutations in *DOCK6*, which encodes a regulator of Rho GTPase signaling, or in *EOGT*, which encodes a component of the Notch pathway, lead to autosomal-recessive forms (Lehman et al. 2014; Shaheen et al. 2011, 2013; Sukalo et al. 2015a, b), mutations in *NOTCH1*, *RBPJ*, or *DLL4*, all of which are Notch pathway components, or in *ARHGAP31*, which encodes another Rho GTPase regulator, result in dominant forms (Table 9.1, Hassed et al. 2012; Israe et al. 2014; Meester et al. 2015; Southgate et al. 2011, 2015; Stittrich et al. 2014). Some studies indicate that DLL4 mutations are more randomly distributed in the ligand, even including two truncation mutations of the C-terminal domain (reviewed in Mašek and Andersson 2017;

Meester et al. 2019). NOTCH1 mutations are in most cases missense mutations in cysteines, predominantly in the ligand-binding domain of EGF-like repeat 11 (reviewed in Mašek and Andersson 2017; Meester et al. 2019).

Clinical Findings in Adams-Oliver Syndrome

Adams-Oliver syndrome is diagnosed based on the presence of several clinical hallmarks (Table 9.1), namely, (a) terminal transverse limb malformations, (b) a local absence of skin (named aplasia cutis congenita, Fig. 9.1), and (c) a partial absence of skull bones. Terminal transverse limb defects can resemble amputations, and patients may also have syndactyly (Table 9.1, reviewed in Mašek and Andersson 2017; Meester et al. 2019; Zanotti and Canalis 2016). In general, aplasia cutis congenita (Fig. 9.1) is predominantly found in the skull region; however other body parts, including the abdomen, may also be affected (reviewed in Meester et al. 2019; Zanotti and Canalis 2016). The severity and symptoms of aplasia cutis congenita and of skull symptoms may greatly vary (reviewed in Meester et al. 2019; Zanotti and Canalis 2016). At birth, the affected skin region typically presents as healed but scarred skin, and skin histology shows absent epidermis, dermal atrophy, and a lack of elastic fibers and other skin structures (reviewed in Meester et al. 2019; Zanotti and Canalis 2016). However, symptoms may range from a localized region with complete absence of skin to patches

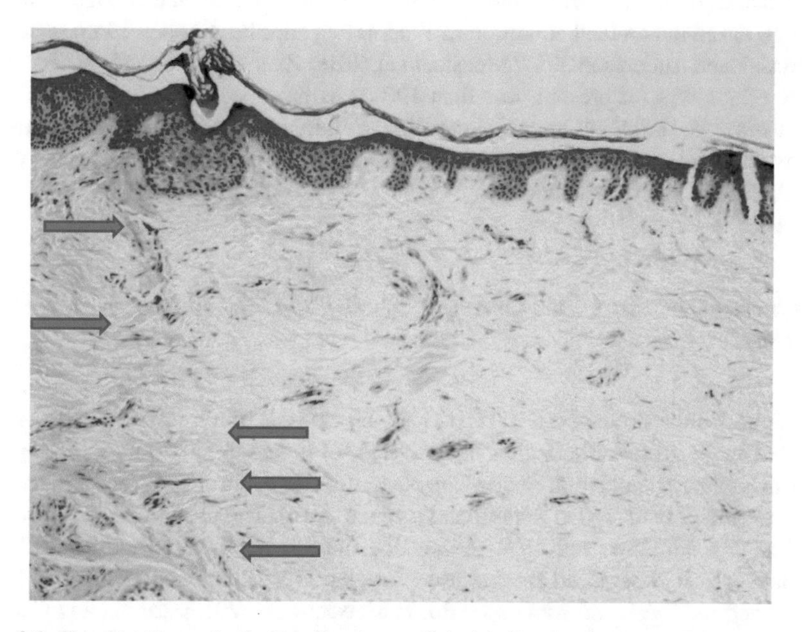

Fig. 9.1 Skin histology of aplasia congenita, a clinical hallmark of Adams-Oliver syndrome. Please note dermal atrophy with lack of elastic fibers and rarefication of epidermal appendages and other skin structures (arrows)

of skin that lack hair (reviewed in Meester et al. 2019; Zanotti and Canalis 2016). Similarly, the many facettes of skull symptoms may range from an absence of skull to a near-normal skull (Lehman et al. 1993; reviewed in Meester et al. 2019; Zanotti and Canalis 2016). Moreover, affected patients may present with vascular anomalies, including dilated surface blood vessels, which result in a marbled appearance of affected skin areas (termed cutis marmorata telangiectatica), pulmonary or portal hypertension, and retinal hypervascularization; around 23% have congenital heart defects (around 23%) (reviewed in Mašek and Andersson 2017; Meester et al. 2019; Zanotti and Canalis 2016). It has been speculated that most symptoms of Adams-Oliver syndrome may be caused by impaired circulation (Patel et al. 2004; Stittrich et al. 2014; Swartz et al. 1999).

Hajdu-Cheney Syndrome (HCS) (OMIN 102500)

Historical Considerations

Hajdu-Cheney syndrome (HCS) represents a devastating disease that was first described in 1948, when Hajdu and Kauntze reported the case of a 37-year-old accountant suffering from severe osteoporosis, acroosteolysis, and neurological complications, who died 12 years later (Hajdu and Kauntze 1948; reviewed in Mašek and Andersson 2017; Meester et al. 2019; Zanotti and Canalis 2016). Cheney then reported the clinical condition in 1965 as a syndrome (Cheney 1965; reviewed in Mašek and Andersson 2017; Meester et al. 2019; Zanotti and Canalis 2016). HCS represents a rare (at present, less than 100 cases have been reported, although its prevalence is probably higher) autosomal-dominant inherited disease although many sporadic cases occur (reviewed in Mašek and Andersson 2017; Meester et al. 2019; Zanotti and Canalis 2016).

Molecular Biology and Genetics of Hajdu-Cheney Syndrome (HCS)

The intracellular domain of NOTCH2 (NICD) consists of a transcriptional domain formed by an Rbpjk association module (RAM) linked to ankyrin (ANK) repeats and a nuclear localization sequence (NLS) (reviewed in Mašek and Andersson 2017; Meester et al. 2019; Zanotti and Canalis 2016). The C-terminus contains the proline (P)-, glutamic acid (E)-, serine (S)-, and threonine (T)-rich motifs (PEST) domain which is required for the ubiquitinylation and degradation of the NICD (reviewed in Mašek and Andersson 2017; Meester et al. 2019; Zanotti and Canalis 2016). More than half a century after the original description, whole-exome sequencing in individuals affected with HCS revealed the presence of point mutations in exon 34 of NOTCH2 leading to the creation of a stop codon and the premature termination of the protein product upstream the PEST domain (Table 9.1, Isidor et al. 2011a, b; Simpson et al. 2011; Majewski et al. 2011). There is now general consensus that these nonsense mutations in exon 34, which lead to the formation of a truncated protein (consisting of all NOTCH2 sequences necessary for the formation of the transcriptional complex, but lacking the PEST domain needed for the ubiquitinylation and degradation of NOTCH2), cause HCS. As a result, the NOTCH2 protein that is synthesized is stable and active. It is of interest that NOTCH2 transcript levels were equivalent to those observed in controls, indicating a reduced capacity to activate the process of nonsense-mediated mRNA decay (reviewed in Mašek and Andersson 2017; Meester et al. 2019; Zanotti and Canalis 2016). Since the PEST domain contains sequences necessary for the ubiquitinylation and degradation of Notch in the proteasome, the mutations lead to a stable protein and persistence of NOTCH2 signaling since all sequences required for the formation of the Notch transcriptional complex are upstream the PEST domain and are therefore preserved characterized by focal bone lysis of distal phalanges and by generalized osteoporosis (Hajdu and Kauntze 1948; Cheney 1965; Silverman et al. 1974; Currarino 2009; Gray et al. 2012).

Hajdu-Cheney and Alagille syndrome can be described as two sides of the same coin. Alagille syndrome is caused by haploinsufficiency (HI) for *JAG1* (~94% of cases) (Li et al. 1997; Oda et al. 1997) or by mutations in *NOTCH2* (~2% of patients) (McDaniell et al. 2006) and is considered to be a Notch loss-of-function phenotype. By contrast, Hajdu-Cheney syndrome, which is driven by production of a stabilized NOTCH2 lacking a functional PEST degradation domain, is caused by gain-of-function mutations in *NOTCH2* (Gray et al. 2012; Han et al. 2015; Isidor et al. 2011a, b; Majewski et al. 2011; Simpson et al. 2011). As we highlight below, a vast number of tissues and organs affected in these syndromes are likely a reflection of the varied and indispensable roles – as revealed by various in vitro studies and knockout studies in animal models – of *Jag1* and *Notch2* in developmental processes (reviewed in Mašek and Andersson 2017; Meester et al. 2019; Zanotti and Canalis 2016).

Clinical Findings in Hajdu-Cheney Syndrome (HCS)

In general, patients suffering from HCS present with abnormalities of craniofacial development that become evident at a young age in childhood and evolve as the person matures (Table 9.1). As in other diseases linked to the Notch pathway, there are high clinical variability and a phenotypical evolution of the disease. HCS is typically associated with facial dysmorphism, synophrys, and epicanthal folds (Table 9.1, reviewed in Mašek and Andersson 2017; Meester et al. 2019; Zanotti and Canalis 2016). In general, the distance between the eyes is relatively short; additionally, thick eyebrows extend toward the midline; malar hypoplasia, a long and smooth philtrum, and micrognathia may be present (reviewed in Canalis 2018; Descartes et al. 2014). Additional clinical signs may include a flattened nasal bridge that then becomes broad, and facial features are coarse and a short neck (reviewed in Mašek and Andersson 2017; Meester et al. 2019; Zanotti and Canalis 2016). The craniofacial developmental defects that have been reported to be present in HCS may include wormian bones, open sutures, platybasia, and basilar invagination (reviewed in Canalis 2018; Mašek and Andersson 2017; Meester et al. 2019; Zanotti and Canalis 2016). It has to be noted that these conditions represent serious manifestations of the disease because they can cause severe neurological complications, which may even result in central respiratory arrest and sudden death (reviewed in Canalis 2018). Abnormal dental eruptions and tooth decay with premature loss of teeth have also often been described in HCS (reviewed in Canalis 2018; Canalis and Zanotti 2014). Additional clinical signs may include short stature, generalized and local joint hypermobility, and vertebral abnormalities (including fractures, kyphosis and scoliosis, and long-bone deformities) (reviewed in Mašek and Andersson 2017; Meester et al. 2019; Zanotti and Canalis 2016). It has to be noted that only a limited number of bone biopsy histologies have been described in the scientific literature, revealing inconclusive results that included normal, increased, and decreased bone remodeling (reviewed in Canalis 2018; Elias et al. 1978; Nunziata et al. 1990; Udell et al. 1986). In selected individuals, an increased number of osteoclasts were found suggesting that enhanced bone resorption is responsible for the bone loss (reviewed in Canalis 2018; Udell et al. 1986). A characteristic feature of HCS is acroosteolysis of the distal phalanges of fingers and toes (reviewed in Mašek and Andersson 2017; Meester et al. 2019; Zanotti and Canalis 2016). This finding is associated with a local inflammatory reaction, pain, and swelling, a process that may cause the loss of the distal phalanges, thereby shortening hands and feet (Table 9.1, reviewed in Mašek and Andersson 2017; Meester et al. 2019; Zanotti and Canalis 2016). Histological analysis has shown an inflammatory process, neovascularization, and fibrosis; however the underlying mechanisms are still not completely understood (reviewed in Mašek and Andersson 2017; Meester et al. 2019; Zanotti and Canalis 2016). It was shown that patients with HCS often suffer from splenomegaly, and effects of Notch2 on B-cell allocation may be of relevance (Adami et al. 2016). As already mentioned above, despite the pronounced skeletal abnormalities reported in HCS, little is known regarding the mechanisms underlying the bone loss. Although the distal phalangeal osteolytic lesions would suggest increased localized bone resorption, there is no information on the mechanisms responsible for the generalized osteoporosis (Canalis et al. 2016; reviewed in Mašek and Andersson 2017; Meester et al. 2019; Zanotti and Canalis 2016). The focal osteolysis is accompanied by neovascularization, inflammation, and fibrosis (reviewed in Zanotti and Canalis 2013, Nunziata et al. 1990; Udell et al. 1986; Elias et al. 1978). Iliac crest biopsies have been reported in a small number of cases of HCS and revealed decreased trabecular bone, normal or increased bone remodeling, and normal or decreased bone formation (reviewed in Zanotti and Canalis 2013; Udell et al. 1986; Blumenauer et al. 2002; Brown et al. 1976; Avela et al. 2011). Whether the osteoblast/osteocyte or the osteoclast is the cell responsible for the presumed change in bone turnover has not been established (reviewed in Mašek and Andersson 2017; Meester et al.

2019; Zanotti and Canalis 2016). In osteoclast precursors, Notch2 induces nuclear factor of T-cell 1 transcription and osteoclastogenesis (Fukushima et al. 2008). This effect is exclusively observed with Notch2 and not with Notch1, but whether this mechanism operates in HCS is not known (reviewed in Mašek and Andersson 2017; Meester et al. 2019; Zanotti and Canalis 2016). Notably, the mechanisms that underlie the acroosteolysis in HCS are not considered to be the same as those that cause the generalized bone loss and fractures that may be present in these patients (reviewed in Mašek and Andersson 2017; Meester et al. 2019; Zanotti and Canalis 2016). Additionally, patients with HCS may present with cardiovascular defects (including patent ductus arteriosus, septal defects, and mitral and aortic valve abnormalities associated with valvular insufficiency or stenosis) (reviewed in Canalis 2018; Kaler et al. 1990; Sargin et al. 2013) or polycystic kidneys (about 10% of HCS patients). Interestingly, it has been reported that serpentine fibula-polycystic kidney syndrome may represent the same disorder as HCS (reviewed in Canalis 2018; Gray et al. 2012; Isidor et al. 2011a, b; Majewski et al. 1993; Narumi et al. 2013). Missense mutations in exon 34 of NOTCH2, upstream of sequences encoding for the PEST domain, were detected in patients affected by this disease (reviewed in Zanotti and Canalis 2013; Isidor et al. 2011a, b; Majewski et al. 1993). Bisphosphonate therapy (alendronate and pamidronate) without or together with anabolic therapy (teriparatide) has been tried for the treatment of the skeletal manifestations of patients with HCS, but there is no convincing evidence that either therapy is successful (reviewed in Zanotti and Canalis 2013; Avela et al. 2011; McKiernan 2008; Galli-Tsinopoulou et al. 2012). Although there are only a limited number of patients suffering from HCS, the detection of a cluster of mutations in a single domain of NOTCH2 in individuals with HCS may improve our understanding of the molecular mechanisms that underlie the development of osteoporosis (reviewed in Canalis 2018; Mašek and Andersson 2017; Meester et al. 2019; Zanotti and Canalis 2016). Multiple attempts to uncover genetic variants that contribute to the risk of osteoporosis have been relatively unsuccessful (reviewed in Zanotti and Canalis 2013; Fisher 2011; Richards et al. 2009). Interestingly, bone mineral density and increased risk of osteoporotic fractures were reported to be associated with a distinct allele of JAG1 in seven independent cohorts of Asian and European females (reviewed in Zanotti and Canalis 2013; Kung et al. 2010).

Cerebral Autosomal-Dominant Arteriopathy with Subcortical Infarcts and Leukoencephalopathy (CADASIL) (OMIM 125310)

Heterozygous mutations in *NOTCH3* have been shown to be associated with this inherited autosomal-dominant hereditary disease, which is at present recognized as the most common cause of inherited stroke and vascular cognitive impairment in adults (Table 9.1; Joutel 2011, 2015; reviewed in Meester et al. 2019). To date, the minimum prevalence is calculated to be at least 4 in 100,000, but could be as high as 1 in 10,000 (reviewed in Meester et al. 2019). CADASIL has an age of onset ranging from young to middle-aged adulthood (reviewed in Meester et al. 2019). Large phenotypic variability can be observed both between and within families, without the presence of clear genotype-phenotype correlations (reviewed in Meester et al. 2019). Clinically, CADASIL patients typically suffer from multiple ischemic strokes caused by an arteriopathy that shows breakdown of vascular smooth muscle cells (vSMCs) – a cell population in which NOTCH3 is highly expressed – and vascular dementia (Table 9.1, Joutel et al. 1996). As a diagnostic hallmark of this disease, cerebral white matter lesions can be detected by magnetic resonance imaging (MRI) (Joutel et al. 2000). While earlier investigations have reported a clustering of CADASIL-related *NOTCH3* mutations in exon 4 (Joutel et al. 1997) or in exon 11 (Dotti et al. 2005), more recent studies that considered the mapping of missense mutations, normalized to exon size, demonstrated four "hotspots" for *NOTCH3* mutations. Although Notch3 governs in vSMCs many cellular key processes, including proliferation, maturation, and survival (Domenga et al. 2004; Wang et al. 2012), it has been reported that most CADASIL-related *NOTCH3* mutations do not cause loss of function: they do not abrogate the capacity of the receptor to mediate signaling. This interesting observation can be explained on the molecular level by the finding that at least the majority of CADASIL-causing *NOTCH3* mutations result in deletion or addition of a cysteine residue in the characteristic epidermal growth factor (EGF)-like repeats located in the extracellular domain (ECD) of this transmembrane receptor protein, thereby causing aggregation of the NOTCH3 ECD into extracellular deposits of granular osmiophilic material (GOM) close to the surface of vSMC, which show a progressive degeneration (reviewed in Meester et al. 2019). The presence of GOM can be confirmed histologically after taking a skin biopsy (reviewed in Meester et al. 2019). In the vast majority of patients, brain MRI reveals extensive hyperintensities of the white matter in the periventricular region, external capsule, and anterior part of the temporal lobes, which can be found even 10–15 years before clinical symptoms, like cognitive decline, arise (reviewed in Meester et al. 2019). It has to be emphasized that the presence of subcortical infarcts and leukoencephalopathy are pathognomonic for the diagnosis of CADASIL (reviewed in Meester et al. 2019). It has been concluded that the pathophysiological mechanism, which causes the aggregation of GOM in the extracellular space close to the surface of vSMC, indicates that *NOTCH3* mutations linked to CADASIL do not cause loss of function, but rather exert neomorphic or toxic effects. This concept is also supported by the surprising clinical finding that patients suffering from homozygous CADASIL mutations experience similar or only slightly more severe symptoms as compared with patients harboring heterozygous mutations (Abou Al-Shaar et al. 2016; Liem et al. 2008; Pippucci et al. 2015; Ragno et al. 2013; Soong et al. 2013; Tuominen et al. 2001; Vinciguerra et al. 2014). It has to be noted that if these mutations were associated with loss of function of Notch3, more severe clinical symptoms should be found in homozygous patients.

Early-Onset Arteriopathy with Cavitating Leukodystrophy (EACL)

Recent investigations further underline the relevance of NOTCH3 deficiency for the pathogenesis of vascular leukoencephalopathies (Table 9.1, reviewed in Mašek and Andersson 2017; Meester et al. 2019; Zanotti and Canalis 2016). While neomorphic heterozygous mutations in this gene cause CADASIL, hypomorphic heterozygous alleles have been occasionally reported to be associated with a spectrum of cerebrovascular phenotypes overlapping CADASIL; however their pathogenic potential is still not completely understood (Pippucci et al. 2015). Pippucci et al. reported a patient with childhood-onset arteriopathy, cavitating leukoencephalopathy with cerebral white matter abnormalities presented as diffuse cavitations, multiple lacunar infarctions, and disseminated microbleeds (Pippucci et al. 2015). They identified a novel homozygous c.C2898A (p.C966*) null mutation in NOTCH3 abolishing NOTCH3 expression and causing NOTCH3 signaling impairment (Pippucci et al. 2015). Several NOTCH3 targets, which either exert their effects in the regulation of arterial tone (KCNA5) or that are expressed in the vasculature (CDH6), were downregulated (Pippucci et al. 2015). Patient's vessels were characterized by smooth muscle degeneration as in CADASIL, but without deposition of granular osmiophilic material (GOM), the CADASIL hallmark (Pippucci et al. 2015). Interestingly, the heterozygous showed similar but less pronounced trends in reduced expression of NOTCH3 and its targets, as well as in vessel degeneration (Pippucci et al. 2015). This study suggests a functional link between NOTCH3 deficiency and pathogenesis of vascular leukoencephalopathies (Pippucci et al. 2015).

Lateral Meningocele Syndrome (LMS) (OMIM 130720)

Lateral meningocele syndrome (LMS), which has also been named Lehman syndrome, represents a very rare inherited disease that has phenotypic overlap with Hajdu-Cheney syndrome (Table 9.1, Gripp et al. 2015). Interestingly, heterozygous truncating *NOTCH3* mutations have been described to be associated with LMS (Gripp et al. 2015). In a recent study, all mutations identified cluster into the last coding exon, resulting in premature termination of the protein and truncation of the negative regulatory proline-glutamate-serine-threonine-rich PEST domain (Gripp et al. 2015). These findings indicate that mutant mRNA products escape nonsense-mediated decay (Gripp et al. 2015). It has been hypothesized that the truncated NOTCH3 may cause gain of function through decreased clearance of the active intracellular product, resembling NOTCH2 mutations in the clinically related Hajdu-Cheney syndrome and contrasting the NOTCH3 missense mutations causing CADASIL (Gripp et al. 2015). Clinically, LMS presents as a skeletal disorder associated with facial anomalies, hypotonia, and/or meningocele-related neurologic malfunction (Gripp et al. 2015). The characteristic lateral meningoceles are the clinical hallmarks of the disease (Gripp et al. 2015). Being in general most severe in the lower spine, they represent the severe end of the dural ectasia spectrum (Gripp et al. 2015). Additional clinical features of LMS include facial abnormality (hypertelorism and telecanthus, high-arched eyebrows, ptosis, midfacial hypoplasia, micrognathia, high and narrow palate, low-set ears, and a hypotonic appearance) and connective tissue abnormalities (hyperextensibility, hernias, and scoliosis) (Gripp et al. 2015). Additionally, aortic dilation, a high-pitched nasal voice, wormian bones, and osteolysis may be found (Gripp et al. 2015).

Recently, a mouse model of LMS (Notch3$^{tm1.1Ecan}$) was established by introducing a tandem termination codon in the Notch3 locus upstream of the proline (P), glutamic acid (E), serine (S), and threonine (T) (PEST) domain (Table 9.1, Yu et al. 2019). Microcomputed tomography showed that these Notch3$^{tm1.1Ecan}$ mice develop osteopenia (Yu et al. 2019). Interestingly, it was shown that the cancellous bone osteopenia was no longer detected after the intraperitoneal administration of antibodies directed against the negative regulatory region (NRR) of Notch3. In that study, the anti-Notch3 NRR antibody suppressed the expression of Hes1, Hey1, and Hey2 (Notch target genes) and decreased Tnfsf11 (receptor activator of NF kappa B ligand) messenger RNA in Notch3$^{tm1.1Ecan}$ osteoblast (OB) cultures (Yu et al. 2019). It was shown that bone marrow-derived macrophages (BMMs) from Notch3$^{tm1.1Ecan}$ mutants exerted enhanced osteoclastogenesis in culture, and this was increased in cocultures with Notch3$^{tm1.1Ecan}$ OB. Moreover, osteoclastogenesis was inhibited by anti-Notch3 NRR antibodies in Notch3$^{tm1.1Ecan}$ OB/BMM cocultures. In conclusion, the results of this study indicate that cancellous bone osteopenia of Notch3$^{tm1.1Ecan}$ mutants can be reversed by anti-Notch3 NRR antibodies, thereby opening new avenues for the treatment of bone osteopenia in patients suffering from LMS (Yu et al. 2019).

Infantile Myofibromatosis-2 (IM-2) (OMIM 615293)

Mutations in the PDGFRB and NOTCH3 genes were recently identified in patients with IM (Table 9.1, Martignetti et al. 2013; reviewed in Mašek and Andersson 2017). While most cases of IM appear to be sporadic, occurrence of IM within families across multiple generations indicates an autosomal-dominant (AD) inheritance pattern; however autosomal-recessive (AR) modes of inheritance have also been suggested. In 2013, Martignetti et al. published an investigation, in which they performed whole-exome sequencing (WES) in members of nine unrelated families clinically diagnosed with AD IM to identify the genetic origin of the disorder (Martignetti et al. 2013). In eight of the families, they detected one of two disease-causing mutations, c.1978C>A (p.Pro660Thr) and c.1681C>T (p.Arg561Cys), in PDGFRB. Interestingly, one family did not have either of these PDGFRB mutations on chromosome 5q32, but all affected individuals had a c.4556T>C (p.Leu1519Pro) mutation in NOTCH3 (Martignetti et al. 2013). In summary, this report indicates two separate forms of IM, namely, IM-1 (OMIM 228550) that is caused by mutations in PDGFRB, and a second form of infantile myofibromatosis, infantile myomatosis-2 (IM-2), that is caused by heterozygous mutation in the NOTCH3 gene on chromosome 19p13. Until today, only one such family has been reported.

Clinically, IM is characterized by nonmetastasizing mesenchymal tumors in the skin, muscle, bone, and viscera, which in most cases develop in neonates or infants under 24 months of age, with few reports of adult onset (reviewed in Mašek and Andersson 2017). Most patients typically present with a single or multiple subcutaneous swellings. Solitary, multicentric or generalized forms of the disease have been reported. Most patients present with a single cutaneous nodule (solitary form). The multicentric is characterized by mesenchymal tumors in the skin, subcutaneous tissues, muscles, and bone. The prognosis is in general good, with no metastases and in many cases regression of the tumor over a period of 12–18 months. In contrast, the generalized form has an unfavorable course, being associated with visceral involvement and having a 76% mortality from cardiopulmonary or gastrointestinal complications. There is no standard therapy and treatment options vary widely. Solitary and even multicentric cutaneous and subcutaneous lesions without visceral involvement often regress spontaneously. However, calcification and atrophic scars frequently remain after spontaneous regression of the lesions. Extensive surgery and chemotherapy were reported to be beneficial for multicentric disease. It has been concluded that further studies of the crosstalk between PDGFRB and NOTCH pathways may offer new opportunities to identify mutations in other genes that result in IM and are a necessary first step toward understanding the mechanisms of both tumor growth and regression and its targeted treatment (Martignetti et al. 2013).

Conclusions

The evolutionary highly conserved Notch pathway regulates many cellular core processes including cell fate decisions and represents one of the most important pathways that govern embryogenesis. Therefore it is no surprise that a broad variety of independent inherited diseases has now been identified (including Alagille, Adams-Oliver, and Hajdu-Cheney syndromes, CADASIL, EACL, LMS, and IM), which are linked to defective Notch signaling. During the last decades, a huge mountain of new scientific information has greatly increased our understanding how Notch governs cell fate decisions and other cellular core processes, thereby not only unraveling hidden secrets concerning the highly orchestrated regulation of embryogenesis but also opening new avenues for the clinical management of Notch-linked inherited diseases.

References

Abou Al-Shaar H, Qadi N, Al-Hamed MH, Meyer BF, Bohlega S (2016) Phenotypic comparison of individuals with homozygous or heterozygous mutation of NOTCH3 in a large CADASIL family. J Neurol Sci 367:239–243. https://doi.org/10.1016/j.jns.2016.05.061

Adami G, Rossini M, Gatti D, Orsolini G, Idolazzi L, Viapiana O et al (2016) Hajdu Cheney Syndrome; report of a novel NOTCH2 mutation and treatment with denosumab. Bone 92:150–156

Adams FH, Oliver CP (1945) Hereditary deformities in man. J Hered 36:3–7. https://doi.org/10.1093/oxfordjournals.jhered.a105415

Alagille D, Odièvre M, Gautier M, Dommergues JP (1975) Hepatic ductular hypoplasia associated with characteristic facies, vertebral malformations, retarded physical, mental, and sexual development, and cardiac murmur. J Pediatr 86(1):63–71. https://doi.org/10.1016/S0022-3476(75)80706-2. PMID 803282

Alagille D, Estrada A, Hadchouel M, Gautier M, Odievre M, Dommergues JP (1987) Syndromic paucity of interlobular bile ducts (Alagille syndrome or arteriohepatic dysplasia): review of 80 cases. J Pediatr 110(2):195–200

Andersson ER, Sandberg R, Lendahl U (2011) Notch signaling: simplicity in design, versatility in function. Development 138:3593–3612. https://doi.org/10.1242/dev.063610

Avela K, Valanne L, Helenius I, Makitie O (2011) Hajdu-Cheney syndrome with severe dural ectasia. Am J Med Genet A 155A:595–598

Blumenauer BT, Cranney AB, Goldstein R (2002) Acro-osteolysis and osteoporosis as manifestations of the Hajdu-Cheney syndrome. Clin Exp Rheumatol 20:574–575

Bosse K, Hans CP, Zhao N et al (2013) Endothelial nitric oxide signaling regulates Notch1 in aortic valve disease. J Mol Cell Cardiol 60:27–35

Bray SJ (2016) Notch signalling in context. Nat Rev Mol Cell Biol 17:722–735. https://doi.org/10.1038/nrm.2016.94

Brown DM, Bradford DS, Gorlin RJ, Desnick RJ, Langer LO, Jowsey J, Sauk JJ (1976) The acro-osteolysis syndrome: morphologic and biochemical studies. J Pediatr 88:573–580

Canalis E (2018) Clinical and experimental aspects of notch receptor signaling: Hajdu-Cheney syndrome and related disorders. Metab Clin Exp 80:48–56. https://doi.org/10.1016/j.metabol.2017.08.002

Canalis E, Zanotti S (2014) Hajdu-Cheney syndrome: a review. Orphanet J Rare Dis 9:200

Canalis E, Schilling L, Yee S-P, Lee S-K, Zanotti S (2016) Hajdu Cheney mouse mutants exhibit osteopenia, increased osteoclastogenesis, and bone resorption. J Biol Chem 291:1538–1551. https://doi.org/10.1074/jbc.M115.685453

Chen J, Ryzhova LM, Sewell-Loftin MK et al (2015) Notch1 mutation leads to valvular calcification through enhanced myofibroblast mechanotransduction. Arterioscler Thromb Vasc Biol 35(7):1597–1605

Cheney WD (1965) Acro-osteolysis. Am J Roentgenol Radium Therapy, Nucl Med 94:595–607

Currarino G (2009) Hajdu-Cheney syndrome associated with serpentine fibulae and polycystic kidney disease. Pediatr Radiol 39:47–52

Descartes M, Rojnueangnit K, Cole L, Sutton A, Morgan SL, Patry L et al (2014) Hajdu-Cheney syndrome: phenotypical progression with de-novo NOTCH2 mutation. Clin Dysmorphol 23:88–94

Dexter JS (1914) The analysis of a case of continuous variation in Drosophila by a study of its linkage relations. Am Nat 48:712–758. https://doi.org/10.1086/279446

Domenga V, Fardoux P, Lacombe P, Monet M, Maciazek J, Krebs LT, Klonjkowski B, Berrou E, Mericskay M, Li Z et al (2004) Notch3 is required for arterial identity and maturation of vascular smooth muscle cells. Genes Dev 18:2730–2735. https://doi.org/10.1101/gad.308904

Dotti MT, Federico A, Mazzei R, Bianchi S, Scali O, Conforti FL, Sprovieri T, Guidetti D, Aguglia U, Consoli D et al (2005) The spectrum of Notch3 mutations in 28 Italian CADASIL families. J Neurol Neurosurg Psychiatry 76:736–738. https://doi.org/10.1136/jnnp.2004.048207

Elias AN, Pinals RS, Anderson HC, Gould LV, Streeten DH (1978) Hereditary osteodysplasia with acro-osteolysis. (The Hajdu-Cheney syndrome). Am J Med 65:627–636

Emerick KM, Rand EB, Goldmuntz E, Krantz ID, Spinner NB, Piccoli DA (1999) Features of Alagille syndrome in 92 patients: frequency and relation to prognosis. Hepatology 29:822–829. https://doi.org/10.1002/hep.510290331

Fisher E (2011) A step forward on the path towards understanding osteoporosis. Clin Genet 80:136–137

Fukushima H, Nakao A, Okamoto F, Shin M, Kajiya H, Sakano S, Bigas A, Jimi E, Okabe K (2008) The association of Notch2 and NF-kappaB accelerates RANKL-induced osteoclastogenesis. Mol Cell Biol 28:6402–6412

Galli-Tsinopoulou A, Kyrgios I, Giza S, Giannopoulou EM, Maggana I, Laliotis N (2012) Two-year cyclic infusion of pamidronate improves bone mass density and eliminates risk of fractures in a girl with osteoporosis due to Hajdu-Cheney syndrome. Minerva Endocrinol 37:283–289

Garg V, Muth AN, Ransom JF, Schluterman MK, Barnes R, King IN, Grossfeld PD, Srivastava D (2005) Mutations in NOTCH1 cause aortic valve disease. Nature 437:270–274. https://doi.org/10.1038/nature03940

Gazave E, Lapébie P, Richards GS, Brunet F, Ereskovsky AV, Degnan BM, Borchiellini C, Vervoort M, Renard E (2009) Origin and evolution of the Notch signalling pathway: an overview from eukaryotic genomes. BMC Evol Biol 9:249. https://doi.org/10.1186/1471-2148-9-249

Gillis E, Kumar AA, Luyckx I et al (2017) Candidate gene resequencing in a large bicuspid aortic valve-associated thoracic aortic aneurysm cohort: SMAD6 as an important contributor. Front Physiol 8:400

Gray MJ, Kim CA, Bertola DR, Arantes PR, Stewart H, Simpson MA, Irving MD, Robertson SP (2012) Serpentine fibula polycystic kidney syndrome is part of the phenotypic spectrum of Hajdu-Cheney syndrome. Eur J Hum Genet 20:122–124

Gridley T (2003) Notch signaling and inherited disease syndromes. Hum Mol Genet 12:9R–13R. https://doi.org/10.1093/hmg/ddg052

Gripp KW, Robbins KM, Sobreira NL, Witmer PD, Bird LM, Avela K, Makitie O, Alves D, Hogue JS, Zackai EH et al (2015) Truncating mutations in the last exon of NOTCH3 cause lateral meningocele syndrome. Am J Med Genet A 167:271–281. https://doi.org/10.1002/ajmg.a.36863

Gunadi, Kaneshiro M, Okamoto T, Sonoda M, Ogawa E, Okajima H, Uemoto S (2019) Outcomes of liver transplantation for Alagille syndrome after Kasai portoenterostomy: Alagille Syndrome with agenesis of extrahepatic bile ducts at porta hepatis. J Pediatr Surg:pii: S0022-3468(19)30323-9. https://doi.org/10.1016/j.jpedsurg.2019.04.022. [Epub ahead of print] PubMed PMID: 31104835

Hajdu N, Kauntze R (1948) Cranio-skeletal dysplasia. Br J Radiol 21:42–48

Han MS, Ko JM, Cho T-J, Park W-Y, Cheong HI (2015) A novel NOTCH2 mutation identified in a Korean family with Hajdu-Cheney syndrome showing phenotypic diversity. Ann Clin Lab Sci 45:110–114

Hansson EM, Lanner F, Das D, Mutvei A, Marklund U, Ericson J, Farnebo F, Stumm G, Stenmark H, Andersson ER et al (2010) Control of Notch-ligand endocytosis by ligand-receptor interaction. J Cell Sci 123:2931–2942. https://doi.org/10.1242/jcs.073239

Hassed SJ, Wiley GB, Wang S, Lee J-Y, Li S, Xu W, Zhao ZJ, Mulvihill JJ, Robertson J, Warner J et al (2012) RBPJ mutations identified in two families affected by Adams-Oliver syndrome. Am J Hum Genet 91:391–395. https://doi.org/10.1016/j.ajhg.2012.07.005

Isidor B, Lindenbaum P, Pichon O, Bézieau S, Dina C, Jacquemont S, Martin-Coignard D, Thauvin-Robinet C, Le Merrer M, Mandel J-L et al (2011a) Truncating mutations in the last exon of NOTCH2 cause a rare skeletal disorder with osteoporosis. Nat Genet 43:306–308. https://doi.org/10.1038/ng.778

Isidor B, Le MM, Exner GU, Pichon O, Thierry G, Guiochon-Mantel A et al (2011b) Serpentine fibula-polycystic kidney syndrome caused by truncating mutations in NOTCH2. Hum Mutat 32:1239–1242

Isrie M, Wuyts W, Van Esch H, Devriendt K (2014) Isolated terminal limb reduction defects: Extending the clinical spectrum of Adams-Oliver syndrome and ARHGAP31 mutations. Am J Med Genet A 164:1576–1579. https://doi.org/10.1002/ajmg.a.36486

Joutel A (2011) Pathogenesis of CADASIL. BioEssays 33:73–80. https://doi.org/10.1002/bies.201000093

Joutel A (2015) The NOTCH3^{ECD} cascade hypothesis of cerebral autosomal-dominant arteriopathy with subcortical infarcts and leukoencephalopathy disease. Neurol Clin Neurosci 3:1–6. https://doi.org/10.1111/ncn3.135

Joutel A, Corpechot C, Ducros A, Vahedi K, Chabriat H, Mouton P, Alamowitch S, Domenga V, Cécillion M, Maréchal E et al (1996) Notch3 mutations in CADASIL, a hereditary adult-onset condition causing stroke and dementia. Nature 383:707–710. https://doi.org/10.1038/383707a0

Joutel A, Vahedi K, Corpechot C, Troesch A, Chabriat H, Vayssière C, Cruaud C, Maciazek J, Weissenbach J, Bousser M-G et al (1997) Strong clustering and stereotyped nature of Notch3 mutations in CADASIL patients. Lancet 350:1511–1515. https://doi.org/10.1016/S0140-6736(97)08083-5

Joutel A, Andreux F, Gaulis S, Domenga V, Cecillon M, Battail N, Piga N, Chapon F, Godfrain C, Tournier-Lasserve E et al (2000) The ectodomain of the Notch3 receptor accumulates within the cerebrovasculature of CADASIL patients. J Clin Invest 105:597–605. https://doi.org/10.1172/JCI8047

Kaler SG, Geggel RL, Sadeghi-Nejad A (1990) Hajdu-Cheney syndrome associated with severe cardiac valvular and conduction disease. Dysmorphol Clin Genet 4:43–47

Kamath BM, Spinner NB, Emerick KM, Chudley AE, Booth C, Piccoli DA, Krantz ID (2004) Vascular anomalies in Alagille Syndrome: a significant cause of morbidity and mortality. Circulation 109:1354–1358. https://doi.org/10.1161/01.CIR.0000121361.01862.A4

Kamath BM, Bauer RC, Loomes KM, Chao G, Gerfen J, Hutchinson A, Hardikar W, Hirschfield G, Jara P, Krantz ID, Lapunzina P, Leonard L, Ling S, Ng VL, Hoang PL, Piccoli DA, Spinner NB (2012) NOTCH2 mutations in Alagille syndrome. J Med Genet 49(2):138–144. https://doi.org/10.1136/jmedgenet-2011-100544. PMC 3682659. PMID 22209762

Kamath BM, Spinner NB, Rosenblum ND (2013) Renal involvement and the role of Notch signalling in Alagille syndrome. Nat Rev Nephrol 9:409–418. https://doi.org/10.1038/nrneph.2013.102

Kerstjens-Frederikse WS, van de Laar IM, Vos YJ et al (2016) Cardiovascular malformations caused by NOTCH1 mutations do not keep left: data on 428 probands with left-sided CHD and their families. Genet Med 18(9):914–923

Kidd S, Kelley MR, Young MW (1986) Sequence of the notch locus of Drosophila melanogaster: relationship of the encoded protein to mammalian clotting and growth factors. Mol Cell Biol 6(9):3094–3108

Kiernan AE, Ahituv N, Fuchs H, Balling R, Avraham KB, Steel KP, Hrabé de Angelis M (2001) The Notch ligand Jagged1 is required for inner ear sensory development. Proc Natl Acad Sci U S A 98:3873–3878. https://doi.org/10.1073/pnas.071496998

Koenig SN, Bosse K, Majumdar U, Bonachea EM, Radtke F, Garg V (2016) Endothelial Notch1 is required for proper development of the semilunar valves and cardiac outflow tract. J Am Heart Assoc 5(4):e003075

Koenig SN, La Haye S, Feller JD et al (2017) Notch1 haploinsufficiency causes ascending aortic aneurysms in mice. JCI Insight 2(21):e91353

Kopan R, Ilagan MXG (2009) The canonical Notch signaling pathway: unfolding the activation mechanism. Cell 137:216–233. https://doi.org/10.1016/j.cell.2009.03.045

Krebs LT, Xue Y, Norton CR et al (2000) Notch signaling is essential for vascular morphogenesis in mice. Genes Dev 14(11):1343–1352

Kung AWC, Xiao S-M, Cherny S, Li GHY, Gao Y, Tso G, Lau KS, Luk KDK, Liu J-M, Cui B et al (2010) Association of JAG1 with bone mineral density and osteoporotic fractures: a genome-wide Association Study and Follow-up Replication Studies. Am J Hum Genet 86:229–239. https://doi.org/10.1016/j.ajhg.2009.12.014

Lee TC, Zhao YD, Courtman DW, Stewart DJ (2000) Abnormal aortic valve development in mice lacking endothelial nitric oxide synthase. Circulation 101(20):2345–2348

Lehman A, Wuyts W, Patel MS (1993) Adams-Oliver syndrome. University of Washington, Seattle

Lehman A, Stittrich A-B, Glusman G, Zong Z, Li H, Eydoux P, Senger C, Lyons C, Roach JC, Patel M (2014) Diffuse angiopathy in Adams-Oliver syndrome associated with truncating DOCK6 mutations. Am J Med Genet A 164:2656–2662. https://doi.org/10.1002/ajmg.a.36685

Li L, Krantz ID, Deng Y, Genin A, Banta AB, Collins CC, Qi M, Trask BJ, Kuo WL, Cochran J et al (1997) Alagille syndrome is caused by mutations in human Jagged1, which encodes a ligand for Notch1. Nat Genet 16:243–251. https://doi.org/10.1038/ng0797-243

Liem MK, Lesnik Oberstein SAJ, Vollebregt MJ, Middelkoop HAM, Grond J, Helderman-van den Enden ATJM (2008) Homozygosity for a NOTCH3 mutation in a 65-year-old CADASIL patient with mild symptoms. J Neurol 255:1978–1980. https://doi.org/10.1007/s00415-009-0036-x

MacGrogan D, D'Amato G, Travisano S, Martinez-Poveda B, Luxán G, Del Monte-Nieto G, Papoutsi T, Sbroggio M, Bou V, Gomez-Del Arco P et al (2016) Sequential ligand-dependent Notch signaling activation regulates valve primordium formation and morphogenesis. Circ Res 118:1480–1497. https://doi.org/10.1161/CIRCRESAHA.115.308077

Majewski F, Enders H, Ranke MB, Voit T (1993) Serpentine fibula--polycystic kidney syndrome and Melnick-Needles syndrome are different disorders. Eur J Pediatr 152:916–921

Majewski J, Schwartzentruber JA, Caqueret A, Patry L, Marcadier J, Fryns JP, Boycott KM, Ste-Marie LG, McKiernan FE, Marik I, Van EH, Michaud JL, Samuels ME (2011) Mutations in NOTCH2 in families with Hajdu-Cheney syndrome. Hum Mutat 32:1114–1117

Martignetti JA, Tian L, Li D, Ramirez MCM, Camacho-Vanegas O, Camacho SC, Guo Y, Zand DJ, Bernstein AM, Masur SK et al (2013) Mutations in PDGFRB cause autosomal-dominant infantile myofibromatosis. Am J Hum Genet 92:1001–1007. https://doi.org/10.1016/j.ajhg.2013.04.024

Mašek J, Andersson ER (2017) The developmental biology of genetic Notch disorders. Development 2017(144):1743–1763. https://doi.org/10.1242/dev.148007

McDaniell R, Warthen DM, Sanchez-Lara PA, Pai A, Krantz ID, Piccoli DA, Spinner NB (2006) NOTCH2 mutations cause Alagille syndrome, a heterogeneous disorder of the notch signaling pathway. Am J Hum Genet 79:169–173. https://doi.org/10.1086/505332

McKiernan FE (2008) Integrated anti-remodeling and anabolic therapy for the osteoporosis of Hajdu-Cheney syndrome: 2-year follow-up. Osteoporos Int 19:379–380

Meester JAN, Southgate L, Stittrich A-B, Venselaar H, Beekmans SJA, den Hollander N, Bijlsma EK, Helderman-van den Enden A, Verheij JBGM, Glusman G et al (2015) Heterozygous loss-of-function mutations in DLL4 cause Adams-Oliver syndrome. Am J Hum Genet 97:475–482. https://doi.org/10.1016/j.ajhg.2015.07.015

Meester JAN, Verstraeten A, Alaerts M, Schepers D, Van Laer L, Loeys BL (2019) Overlapping but distinct roles for NOTCH receptors in human cardiovascular disease. Clin Genet 95(1):85–94. https://doi.org/10.1111/cge.13382. Epub 18 Jun 10. Review. PubMed PMID: 29767458

Michelena HI, Prakash SK, Della Corte A et al (2014) Bicuspid aortic valve: identifying knowledge gaps and rising to the challenge from the international bicuspid aortic valve consortium (BAVCon). Circulation 129(25):2691–2704

Morgan TH (1917) The theory of the gene. Am Nat 19:309–310. https://doi.org/10.1086/279629

Morgan T (1928) The theory of the gene, revised edn. Yale University Press, New Haven, pp 77–81

Narumi Y, Min BJ, Shimizu K, Kazukawa I, Sameshima K, Nakamura K, Kosho T, Rhee Y, Chung YS, Kim OH, Fukushima Y, Park WY, Nishimura G (2013) Clinical consequences in truncating mutations in exon 34 of NOTCH2: report of six patients with Hajdu-Cheney syndrome and a patient with serpentine fibula polycystic kidney syndrome. Am J Med Genet A 161A(3):518–526. https://doi.org/10.1002/ajmg.a.35772. Epub 2013 Feb 7. Erratum in: Am J Med Genet A. 2013;161(10):2685. PubMed PMID: 23401378

Nunziata V, di GG, Ballanti P, Bonucci E (1990) High turnover osteoporosis in acro-osteolysis (Hajdu-Cheney syndrome). J Endocrinol Investig 13:251–255

Oda T, Elkahloun AG, Pike BL, Okajima K, Krantz ID, Genin A, Piccoli DA, Meltzer PS, Spinner NB, Collins FS et al (1997) Mutations in the human Jagged1 gene are responsible for Alagille syndrome. Nat Genet 16:235–242. https://doi.org/10.1038/ng0797-235

Patel MS, Taylor GP, Bharya S, Al-Sanna'a N, Adatia I, Chitayat D, Suzanne Lewis ME, Human DG (2004) Abnormal pericyte recruitment as a cause for pulmonary hypertension in Adams-Oliver syndrome. Am J Med Genet 129A:294–299. https://doi.org/10.1002/ajmg.a.30221

Pippucci T, Maresca A, Magini P, Cenacchi G, Donadio V, Palombo F, Papa V, Incensi A, Gasparre G, Valentino ML et al (2015) Homozygous NOTCH3 null mutation and impaired NOTCH3 signaling in recessive early-onset arteriopathy and cavitating leukoencephalopathy. EMBO Mol Med 7:1–11. https://doi.org/10.15252/emmm.201404399

Ragno M, Pianese L, Morroni M, Cacchiò G, Manca A, Di Marzio F, Silvestri S, Miceli C, Scarcella M, Onofrj M et al (2013) "CADASIL coma" in an Italian homozygous CADASIL patient: comparison with clinical and MRI findings in age-matched heterozygous patients with the same G528C NOTCH3 mutation. Neurol Sci 34:1947–1953. https://doi.org/10.1007/s10072-013-1418-5

Reichrath J, Reichrath S (2020) Notch signalling and Embryonic development: an ancient friend, revisited. Adv Exp Med Biol 1218:9–38

Richards GS, Degnan BM (2009) The dawn of developmental signaling in the metazoa. Cold Spring Harb Symp Quant Biol 74:81–90. https://doi.org/10.1101/sqb.2009.74.028

Richards JB, Kavvoura FK, Rivadeneira F, Styrkarsdottir U, Estrada K, Halldorsson BV, Hsu YH, Zillikens MC, Wilson SG, Mullin BH, Amin N, Aulchenko YS, Cupples LA, Deloukas P, Demissie S, Hofman A, Kong A, Karasik D, van Meurs JB, Oostra BA, Pols HA, Sigurdsson G, Thorsteinsdottir U, Soranzo N, Williams FM, Zhou Y, Ralston SH, Thorleifsson G, Van Duijn CM, Kiel DP, Stefansson K, Uitterlinden AG, Ioannidis JP, Spector TD (2009) Collaborative meta-analysis: associations of 150 candidate genes with osteoporosis and osteoporotic fracture. Ann Intern Med 151:528–537

Sargin G, Cildag S, Senturk T (2013) Hajdu-Cheney syndrome with ventricular septal defect. Kaohsiung J Med Sci 29:343–344

Shaheen R, Faqeih E, Sunker A, Morsy H, Al-Sheddi T, Shamseldin HE, Adly N, Hashem M, Alkuraya FS (2011) Recessive mutations in DOCK6, encoding the guanidine nucleotide exchange factor DOCK6, lead to abnormal actin cytoskeleton organization and Adams-Oliver syndrome. Am J Hum Genet 89:328–333. https://doi.org/10.1016/j.ajhg.2011.07.009

Shaheen R, Aglan M, Keppler-Noreuil K, Faqeih E, Ansari S, Horton K, Ashour A, Zaki MS, Al-Zahrani F, Cueto-González AM et al (2013) Mutations in EOGT confirm the genetic heterogeneity of autosomal-recessive Adams-Oliver Syndrome. Am J Hum Genet 92:598–604. https://doi.org/10.1016/j.ajhg.2013.02.012

Sijmons RH (2008) Encyclopaedia of tumour-associated familial disorders. Part I: from AIMAH to CHIME syndrome. Hered Cancer Clin Pract 6(1):22–57

Silverman FN, Dorst JP, Hajdu N (1974) Acroosteolysis (Hajdu-Cheney syndrome). Birth Defects Orig Artic Ser 10:106–123

Simpson MA, Irving MD, Asilmaz E, Gray MJ, Dafou D, Elmslie FV, Mansour S, Holder SE, Brain CE, Burton BK, Kim KH, Pauli RM, Aftimos S, Stewart H, Kim CA, Holder-Espinasse M, Robertson SP, Drake WM, Trembath RC (2011) Mutations in NOTCH2 cause Hajdu-Cheney syndrome, a disorder of severe and progressive bone loss. Nat Genet 43:303–305

Singh SP, Pati GK (2018) Alagille syndrome and the liver: current insights. Euroasian J Hepatogastroenterol 8(2):140–147

Soong B-W, Liao Y-C, Tu P-H, Tsai P-C, Lee I-H, Chung C-P, Lee Y-C (2013) A homozygous NOTCH3 mutation p.R544C and a heterozygous TREX1 variant p.C99MfsX3 in a family with hereditary small vessel disease of the brain. J Chin Med Assoc 76:319–324. https://doi.org/10.1016/j.jcma.2013.03.002

Southgate L, Sukalo M, Karountzos ASV, Taylor EJ, Collinson CS, Ruddy D, Snape KM, Dallapiccola B, Tolmie JL, Joss S et al (2015) Haploinsufficiency of the NOTCH1 receptor as a cause of Adams-Oliver syndrome with variable cardiac anomalies. Circ Cardiovasc Genet 8:572–581. https://doi.org/10.1161/CIRCGENETICS.115.001086

Southgate L, Machado RD, Snape KM, Primeau M, Dafou D, Ruddy DM, Branney PA, Fisher M, Lee GJ, Simpson MA et al (2011) Gain-of-function mutations of ARHGAP31, a Cdc42/Rac1 GTPase regulator, cause syndromic cutis aplasia and limb anomalies. Am J Hum Genet 88:574–585. https://doi.org/10.1016/j.ajhg.2011.04.013

Stittrich A-B, Lehman A, Bodian DL, Ashworth J, Zong Z, Li H, Lam P, Khromykh A, Iyer RK, Vockley JG et al (2014) Mutations in NOTCH1 cause Adams-Oliver syndrome. Am J Hum Genet 95:275–284. https://doi.org/10.1016/j.ajhg.2014.07.011

Sukalo M, Tilsen F, Kayserili H, Müller D, Tüysüz B, Ruddy DM, Wakeling E, Ørstavik KH, Snape KM, Trembath R et al (2015a) DOCK6 mutations are responsible for a distinct autosomal-recessive variant of Adams-Oliver syndrome associated with brain and eye anomalies. Hum Mutat 36:593–598. https://doi.org/10.1002/humu.22795

Sukalo M, Tilsen F, Kayserili H, Müller D, Tüysüz B, Ruddy DM, Wakeling E, Ørstavik KH, Bramswig NC, Snape KM et al (2015b) DOCK6 mutations are responsible for a distinct autosomal-recessive variant of Adams-Oliver syndrome associated with brain and eye anomalies. Hum Mutat 36:1112–1112. https://doi.org/10.1002/humu.22830

Swartz EN, Sanatani S, Sandor GGS, Schreiber RA (1999) Vascular abnormalities in Adams-Oliver syndrome: cause or effect? Am J Med Genet 82:49–52. https://doi.org/10.1002/(SICI)1096-8628(19990101)82:1<49::AID-AJMG10>3.0.CO;2-M

Tsai H, Hardisty RE, Rhodes C, Kiernan AE, Roby P, Tymowska-Lalanne Z, Mburu P, Rastan S, Hunter AJ, Brown SDM et al (2001) The mouse slalom mutant demonstrates a role for Jagged1 in neuroepithelial patterning in the organ of Corti. Hum Mol Genet 10:507–512. https://doi.org/10.1093/hmg/10.5.507

Tuominen S, Juvonen V, Amberla K, Jolma T, Rinne JO, Tuisku S, Kurki T, Marttila R, Pöyhönen M, Savontaus M-L et al (2001) Phenotype of a homozygous CADASIL patient in comparison to 9 age-matched heterozygous patients with the same R133C Notch3 mutation. Stroke 32:1767–1774. https://doi.org/10.1161/01.STR.32.8.1767

Udell J, Schumacher HR Jr, Kaplan F, Fallon MD (1986) Idiopathic familial acroosteolysis: histomorphometric study of bone and literature review of the Hajdu-Cheney syndrome. Arthritis Rheum 29:1032–1038

Vinciguerra C, Rufa A, Bianchi S, Sperduto A, De Santis M, Malandrini A, Dotti MT, Federico A (2014) Homozygosity and severity of phenotypic presentation in a CADASIL family. Neurol Sci 35:91–93. https://doi.org/10.1007/s10072-013-1580-9

Vrijens K, Thys S, De Jeu MT, Postnov AA, Pfister M, Cox L, Zwijsen A, Van Hoof V, Mueller M, De Clerck NM et al (2006) Ozzy, a Jag1 vestibular mouse mutant, displays characteristics of Alagille syndrome. Neurobiol Dis 24:28–40. https://doi.org/10.1016/j.nbd.2006.04.016

Wang Q, Zhao N, Kennard S, Lilly B (2012) Notch2 and notch3 function together to regulate vascular smooth muscle development. PLoS One 7:e37365. https://doi.org/10.1371/journal.pone.0037365

Wang Y, Wu B, Farrar E et al (2017) Notch-Tnf signalling is required for development and homeostasis of arterial valves. Eur Heart J 38(9):675–686

Wharton KA, Johansen KM, Xu T, Artavanis-Tsakonas S (1985) Nucleotide sequence from the neurogenic locus notch implies a gene product that shares homology with proteins containing EGF-like repeats. Cell 43(3 Pt 2):567–581

Yu J, Siebel CW, Schilling L, Canalis E (2019) An antibody to Notch3 reverses the skeletal phenotype of lateral meningocele syndrome in male mice. J Cell Physiol. https://doi.org/10.1002/jcp.28960. [Epub ahead of print] PubMed PMID: 31188489

Zanotti S, Canalis E (2013) Notch signaling in skeletal health and disease. Eur J Endocrinol 168(6):R95–R103. https://doi.org/10.1530/EJE-13-0115. Print 2013 Jun. Review. PubMed PMID: 23554451; PubMed Central PMCID: PMC4501254

Zanotti S, Canalis E (2016) Notch signaling and the skeleton. Endocr Rev 37:223–253. https://doi.org/10.1210/er.2016-1002

Index

J. Reichrath, S. Reichrath (eds.), *Notch Signaling in Embryology and Cancer*, Advances in Experimental Medicine and Biology 1218,
https://doi.org/10.1007/978-3-030-34436-8

Printed in the United States
By Bookmasters